Technician Class
New No-Code

By
Gordon West
WB6NOA

Master Publishing, Inc.

This book was developed and published by:
Master Publishing, Inc.
Richardson, Texas

Design and artwork by:
Plunk Design
Dallas, Texas

Editing by:
Gerald Luecke

Printing by:
Arby Graphic Service
Chicago, Illinois

Photograph Credit:
All photographs that do not have a source identification are either courtesy of
Radio Shack or Master Publishing, Inc. originals. Appreciation is expressed
to Tom Barnum, AA6TP, and Sarah Spencer for participating in the cover
photograph.

REGARDING THESE BOOK MATERIALS

Table of Contents

QUESTION POOL NOMENCLATURE

The latest nomenclature changes recommended by the volunteer examiner coordinator's Question Pool Committee (QPC) for question pools effective July 1, 1990 have been incorporated in this book.

Our interpretation and implementation may be different from other publishers; therefore, there may be variations in some of the question pool wording. We are sure that any slight differences will not contribute to improper understanding of the question or it's answer.

Preface

Welcome to the fabulous hobby of Amateur Radio! Federal Communications Commission (FCC) action on February 14, 1991, has made it easier to obtain an entry-level license in the ham radio service. In just one test session, you may satisfy the requirements for obtaining the Technician Class license. Absolutely no knowledge of the Morse code is required!

The *New No-Code Technician Class* FCC license preparation book has been specifically designed to help you quickly prepare for your entry-level, license test. The test consists of two simple written examinations, both of which may be administered in the same examination session, one after the other. Element 2 is your 30-question, multiple-choice examination that Novice Class applicants also take, and Element 3A will be your 25-question examination that was the Technician Class examination prior to February 14, 1991. This new revised book contains the 372-question Element 2 pool, effective July 1, 1990, and the 326-question Element 3A pool, also effective July 1, 1990. *Everything* you need to prepare for these two examinations for your Technician Class new no-code license is found right here in this book.

Chapter 1 tells you all about the amateur service, which we affectionately call "ham radio". Chapter 2 shows you all of the exciting bands that await you with full privileges with your new Technician Class license. You'll want to keep this chapter handy when you begin operating because valuable band plan information is given for any frequency on which you may wish to operate. Chapter 3 prepares you for the upcoming two question pools, and shows you the number of questions selected from each subject area.

After you read Chapter 3, jump right into the Element 2 and Element 3A question pools, and start memorizing the material. You'll enjoy the fun we have in telling you all about the answers behind every question! And finally, Chapter 4 tells you exactly what you need to bring when you show up in front of the three accredited volunteer examiners to take your combined written examinations. After the examinations, check with your VEs to make sure your FCC 610 application form is OK. They will fill in, sign, and send to the FCC.

So welcome, again, to Amateur Radio, and we hope to be hearing you on Technician Class frequencies soon. In less than 30 days, you should be ready for the tests. Good luck!

Gordon West, WB6NOA

1

Amateur Service

ABOUT THIS BOOK

The purpose of the *New No-Code Technician Class* book is to prepare you so you can obtain the Technician Class amateur operator license. The license test is simply passing two written examinations, a 30-question Element 2 examination, and a 25-question Element 3A test. Both exams are multiple-choice, and there is no requirement for taking a Morse code test.

We will be talking about the new Technician Class license in detail in the rest of this book. But before we get down to the Technician Class privileges, let's spend a few minutes bringing you up to date on the exciting background of the amateur service, and the requirements of an amateur service license.

WHAT IS THE AMATEUR SERVICE?

More than 450,000 Americans are licensed amateur operators. According to our Federal Communications Commission (FCC), the U.S. Government agency responsible for licensing amateur operators, "The amateur service is for qualified persons of all ages who are interested in radio technique solely with a personal aim and without pecuniary interest." Ham radio, as it is known, is first, and foremost, *a fun hobby!* In addition, it is a service.

The amateur service exists under international treaty and is authorized in practically every country. Each government has its own rules for admission to the ham ranks. Numerous frequency bands throughout the radio spectrum are allocated to the amateur service on an international basis, making it possible for amateur operators to communicate with each other in all parts of the world—even in space. Astronaut Owen Garriott, W5LFL, became the first ham operator to operate from space during the Columbia/Spacelab mission in 1983 using a hand-held radio and an antenna in the window of the space shuttle. Gordon West, your author, was among the first hams to communicate with W5LFL in the Spacelab.

More than 1,500,000 operators exchange ham radio greetings and messages by voice, teleprinting, telegraphy, facsimile and television worldwide. Japan, which has a no-code license, has over one million hams alone! There are 20,000 in the U.S.S.R. where it is considered a national sport. It is very commonplace for U.S. amateurs to communicate with Soviet amateurs, while mainland China is just getting started with their amateur service. Being a ham operator is a very good way to promote international good will.

AMATEUR SERVICE BENEFITS

The benefits of ham radio are countless! There is something for everyone. The ham operators are probably best known for their contributions during times of disaster. When all else fails, the ham operator traditionally gets through. Over the years, amateurs have contributed much to electronic technology. They have even designed and built their own orbiting communications satellites.

Amateur service isn't just for the technically inclined. It is for everyone of all ages. There are ham radio operators under eight years old—and many over eighty! Most hams are just plain folks, but many famous celebrities are amateur operators. Barry Goldwater is K7UGA, Jordan's King Hussein is JY1, ex-pro baseball player Joe Rudi is NK7U, guitarist Chet Atkins is WA4CZD, and singer Ronnie Milsap is WB4KCG. Other famous Amateurs are Bill Halligan (who founded Hallicrafters), Arthur Godfrey, Andy Devine, Pee Wee Hunt, Alvino Rey, General Curtis LeMay . . . and recent Gordon West Radio School graduate, Priscilla Presley.

Ham radio for the handicapped is a godsend! It is a great equalizer, since everyone is the same behind a microphone or a packet radio computer keyboard. Amateur service can take the disabled to the far flung corners of the world. Recent FCC rules for the disabled may now allow special examination techniques to be offered anyone with a handicap. New FCC rules also allow for code test exemptions for General Class and Extra Class with a doctor-certified handicap.

The ham fraternity knows no geographic, political or social barrier. And you are going to be part of it. Probably the primary prerequisite for passing any amateur license examination is the will to do it. If you follow the suggestions in this book, your chances are excellent. If you are fascinated by radio communication, learning will be easy and fun!

LICENSE PRIVILEGES

An amateur operator license conveys many privileges. As the *control operator* of an amateur station, an amateur operator is responsible for the quality of the station's transmissions. Most radio equipment must be authorized by the government before it can be widely used by the public, but for the most part, not amateur equipment! Unlike the citizen's band service, amateurs may design, construct, modify and repair their own equipment.

It is easier than ever to obtain the needed licenses from the FCC; however, there are certain things you must know. It is all covered in this book.

OPERATOR LICENSE REQUIREMENTS

To qualify for a ham license, a person must pass an examination according to FCC guidelines. The degree of skill and knowledge the candidate can demonstrate to the examiners determines the class of operator license for which the person is qualified.

The new rules and regulations no longer require a Morse code test for the entry-level, Technician Class license. While Morse code examinations are still in place for Novice Class (5 wpm), General Class (13 wpm), and Extra class (20 wpm), you don't need to learn the code to get onto those exciting VHF and UHF line-of-sight frequencies plus 6 meters for worldwide communications. Once you're on the air as a Technician, you can think about learning the code at anytime.

There is a big difference between CB and amateur operation. The amateur service is both a public service and a hobby. CB exists for short-distance, low-power, personal and business communications only. There is no 150-mile distance limitation in ham radio. You can talk around the world with higher output power levels—limited only by radio propagation conditions. Hams are licensed with call signs and don't use "handles." Licensing and call signs were discontinued for the citizens band some years ago. CBers are prohibited from experimenting, which is the cornerstone and one of the fun activities of Amateur Radio. Hams can also interconnect their radios with the telephone system. You'll find that ham radio offers you far more capabilities— more frequencies, higher power, more emissions, more modes—and lots of fun, too.

Anyone is eligible to become a U.S. licensed amateur operator (including aliens if they are not a representative of a foreign government). There is no age limitation. If you can pass the exams, you can become a ham!

One of the reasons for the existence of amateur services is to provide communications in times of emergency. Although hardly ever used on the ham bands during an emergency anymore, CW (Morse code) is one way to pierce through interference when other modes cannot get through. Packet radio is another very special type of emergency communications, using your own home computer over the airwaves. Now that there is an opportunity to become an amateur operator Technician Class licensee, you can concentrate on packet communications without ever having to worry about learning the Morse code for emergency communications.

But many amateurs find the code to be fun, and once you have your Technician Class license in hand, you can take any code test of your choice to add more privileges to your new license. And once you've passed a code test, your expanded privileges, with code, will be called Technician Plus.

So don't worry about the Morse code for now—it's something you can learn once you are on the airwaves as a Technician Class operator. There is no pressure to learn the code, and the best news of all, you don't need to learn code to become a Technician Class operator.

OPERATOR LICENSE CLASSES

There are five official successive levels of amateur operator licenses. Each requires progressively higher levels of learning and proficiency, and each gives you additional operating privileges. There is no waiting time to upgrade from one amateur class to another, nor any required waiting time to retake a failed examination. You can take all of the examinations at one sitting if you want to.

This is what is known as *incentive licensing*—a method of strengthening the service by offering more privileges in exchange for more electronic knowledge and code skill. The theory and regulations covered in each of the questions of the various examinations relates to privileges that you will obtain when you upgrade. *Table 1-1* illustrates six levels of operating privileges for five classes of amateur operators. Note that the sixth level is actually an extension of the Technician Class license, the Technician Plus, which is Technician "plus" passing a code test.

Table 1-2 details the subjects covered in the various written examination question elements.

The distribution of the questions in each license class over the topics is shown in *Table 1-3*.

TWO ENTRY-LEVEL CLASSES

There are two entry-level beginner licenses—the traditional *Novice Class operator* requiring a code test, and the *Technician Class New No-Code operator*. If you have been practicing the Morse code, or if you know the sounds of Morse code from the Boy Scouts or the military, then indeed plan to get started in ham radio by taking a code test along with the Novice theory material covered in this book.

However, if you are not interested in learning the Morse code, or have tried to learn the Morse code and just couldn't quite get an ear for it, then indeed join us in the amateur service by going the way of a Technician Class license.

Novice Class for Those That Enjoy Code

If you enjoy the Morse code, join the amateur service by preparing for your Novice Class license. The Novice elementary rules and theory test (Element 2) and its 5-wpm beginner's telegraphy test (Element 1A) make it the ideal license to receive amateur service call letters. The Novice license may be administered by any two General Class hams in the privacy of your own home. To take the Novice test, you do *not* have to go down and appear in front of a team of three accredited

Table 1-1. Amateur License Classes and Exam Requirements

License Class	Test Element	Type of Examination
Novice Class	Element 2	30-Question Written Examination
	Element 1A	5-Words-Per-Minute Code Test
[1,2] Technician Class	Element 2 and 3A	55-Question Written Examination (In 2 parts-30 Element 2, 25 Element 3A) (No Morse code requirement)
[2] Technician Plus Class	Element 3A	25-Question Written Examination if a Novice. 5-wpm Code Test if a Technician.
General Class	Element 3B	25-Question Written Examination
	Element 1B	13-Words-Per-Minute Code Test
Advanced Class	Element 4A	50-Question Written Examination (No additional Morse code requirement)
Extra Class	Element 4B	40-Question Written Examination
	Element 1C	20-Words-Per-Minute Code Test

[1] No-Code License
[2] Effective 2/14/91

Note: Written examinations must be taken in strict ascending order of difficulty all the way to Extra Class. You can't be administered Element 3A until you have passed Element 2, etc. The code tests may be taken in any order. You can take the 20-wpm code test first if you can pass it. You can enter as a Technician without code, and then gain Technician Plus CW privileges by passing the Element 1A code test. You can enter as a Novice with code, and then gain Technician Plus by passing the Element 3A theory examination.

Table 1-2. Question Element Subjects

Element 2
    Novice
    Technician
	Elementary theory and regulations

Element 3A
    Technician
    Technician Plus
	Beginner-Level theory and regulations with VHF/UHF emphasis

Element 3B General	General theory and regulations with emphasis on General Class operating privileges and on HF operation
Element 4A Advanced	Intermediate theory and rules and regulations
Element 4B Extra	Specialized theory and VEC regulations

Note: All license written examinations are additive. For example, to obtain a General Class license, you must take and pass an Element 2 written examination, an Element 3A written examination and an Element 3B written examination. You may not skip over a lower class written examination.

Table 1-3. Topic Distribution Over License Classes

Element		2 Novice Technician	3A Technician Tech Plus	3B General	4A Advanced	4B Extra
A	Commission's Rules	10	5	4	6	8
B	Operating Procedures	2	3	3	1	4
C	Radio Wave Propagation	1	3	3	2	2
D	Amateur Radio Practices	4	4	5	4	4
E	Electrical Principles	4	2	2	10	6
F	Circuit Components	2	2	1	6	4
G	Practical Circuits	2	1	1	10	4
H	Signals and Emissions	2	2	2	6	4
I	Antennas and Feed Lines	3	3	4	5	4
Total questions		30	25	25	50	40

volunteer examiners. If you live miles away from a local accredited volunteer examiner testing team, then go the way of the Novice, and take your test from any two local hams who have a General Class license, or higher.

You will probably find taking the Novice examination from two General Class hams a little bit easier than the more formal examinations from an accredited team of three volunteer examiners (VEs). Most Generals will make up an ultra-simple code test, so if you know a little code, think about starting ham radio as a Novice by using the Element 2 question pool in this book.

Technician Class

Another way to get into the amateur service is via a Technician Class license. This consists of passing written examinations from Element 2 and Element 3A. There is no code test for the Technician Class new no-code license. Both elements are contained in this book, and are easily passed with just a few weeks of going over the questions, the possible right answers, the correct answer, and the fun explanations.

The Technician written examinations must be taken in front of a team of three accredited volunteer examiners. You cannot take the Technician examination in front of two General Class ham buddies— they can only administer Element 1A, Novice code, and Element 2, Novice theory. An interesting fact is that, alternatively, a team of three accredited volunteer examiners can recognize, and give credit for, your having passed either of the Novice Class elements in front of your two General Class ham buddies, if it is properly documented.

The Technician Class new no-code license has been years in the making. The Federal Communications Commission opened up this new way of obtaining a no-code amateur services license on February 14, 1991, after studying numerous petitions and public comments for an easier way to get into Amateur Radio. By offering a codeless class of license with privileges exclusively above 50 MHz, an entry-level license is finally available to those who find the Morse code a barrier to becoming a licensed amateur operator.

But the FCC decided to retain the current Novice Class operator license as an alternate entry-level license for those persons able to pass the 5-wpm Morse code test, instead of the more comprehensive written examinations required for the Technician Class license.

"The amateur service is not growing as it should related to what it has to offer," comments FCC Private Radio Bureau Chief, Ralph Haller at a press conference right after the no-code license was announced. "...The amateur service is where our nation's technical expertise comes from—that the changes should attract people who are interested in computers and digital communications, and should help the U.S. to become more competitive," adds Haller.

By granting the Technician Class operator all amateur privileges from 30 MHz on up, the new Technician Class operator will mainstream easily into the amateur service with little or no negative impact on present VHF and UHF operators. In fact, new Technician Class operators will help fill those bands that are vacant, and save us from possibly losing valuable frequencies in the future. "Use it or lose it" is the motto of amateur operators when looking at the vast expanse of Technician Class frequencies under-utilized throughout the country.

The response to a Technician Class license has been overwhelming. Gordon West Radio School weekend classes have tripled in size, and these classes are now offered almost every weekend somewhere in the country.

Technician Class operators receive group "C" call signs where available, and this further helps them blend in with older operators who originally took a code test to get into the amateur service.

Technician Plus Class

The "plus" added to the Technician Class license means "plus code." Holding a Technician Plus license means you either entered the amateur service by passing, via two ham buddies, a Novice 5-wpm code test and a written theory examination, and then passed a VE-administered Technician written examination—or, the Technician Plus license might mean you entered amateur service by passing two written examinations in front of three accredited volunteer examiners, and then—or at a later date—passed a CW Morse code examination.

For licensed Technician Class operators to earn their Technician Plus license, the optional 5-wpm Morse code test must be taken in front of a team of three accredited examiners. You *are not* permitted to take the 5-wpm code test in front of any two ham buddies with a General Class license or above *once you enter the ham service through the Technician Class new no-code route.*

You may wish to re-read the comments about the Novice Class license; and if you know some code now, reconsider how you plan to enter ham radio. Maybe the best way is via the traditional Novice Class route by taking the Element 2 written theory examination and a code test in front of two General Class ham buddies. After accomplishing that step, go to a VE session, and pass the Element 3A written examination. You are then a Technician Plus operator.

The big advantage of the Technician Plus endorsement are Novice Class code and voice operating privileges *below* 30 MHz. This includes the popular voice portion on the 10-meter band, and the CW Novice privileges on 15 meters, 40 meters, and 80 meters. If you've heard about all the excitement on the 10-meter band for long-range voice communications, study the 5-wpm Radio Shack code tapes from the *New Novice Voice Class* license preparation materials, and ultimately add a "plus" to your Technician Class license. Or, learn code and theory and pass the Novice requirements, then pass the Element 3A written examination, and now you are a Technician Plus operator.

General Class

The third step up the ladder is *General operator*. Another written examination (Element 3B) and proven 13-wpm telegraphy skill (Element 1B) is required. This license authorizes all emission privileges on at least some portions of all amateur service frequency bands, including the very popular DX-oriented 20-meter ham band.

Advanced Class

The fourth step up the ladder is *Advanced operator*. More theory is required (Element 4A). This license authorizes additional frequency privileges on amateur service HF bands. There is no additional code requirement at this step.

Extra Class

About ten percent make it all the way to the top-of-the-line, the *Amateur Extra* Class level. Another written examination (Element 4B) and a 20-wpm telegraphy test (Element 1C) must be passed. This license authorizes all frequency privileges in all communications modes and emissions permitted to the amateur service. It offers you everything there is (or will be as technology develops) and is the ultimate goal of most ham operators.

AMATEUR CALL SIGNS

As an aid to enforcement of the radio rules, transmitting stations throughout the world are required to identify themselves at intervals when they are in operation. By international agreement, the prefix letters of a station's call sign indicates the country in which the station is authorized to operate. On the DX airwaves, hams can readily identify the national origin of the ham signal they hear by its call sign prefix. The national prefixes allocated to the United States are AA through AL, KA through KZ, NA through NZ, and WA through WZ. In addition, U.S. amateur stations with call signs that start with A, K, N or W are followed by a number indicating U.S. geographic area. *Table 1-4* details these geographical areas.

The suffix letters indicate a specific amateur station. The primary Amateur Radio call sign is issued by the FCC's licensing facility in Gettysburg, Pennsylvania, on a systematic basis after they receive your application from the VE team. A call sign is a very important matter to a ham—sometimes more personal than his name! Novice and some Technician call signs contain two prefix letters, a single radio district number (1 through 0) and three suffix letters. KA7ABC is an example.

Table 1-4. Call Sign Numbers for Geographical Areas

Call Sign Area No.	Geographical Area
1	Maine, New Hampshire, Vermont, Massachusetts, Rhode Island, Connecticut.
2	New York, New Jersey.
3	Pennsylvania, Delaware, Maryland, District of Columbia.
4	Virginia, North and South Carolina, Georgia, Florida, Alabama, Tennessee, Kentucky, Puerto Rico and Virgin Islands.
5	Mississippi, Louisiana, Arkansas, Oklahoma, Texas, New Mexico.
6	California, Hawaii.
7	Oregon, Washington, Idaho, Montana, Wyoming, Arizona, Nevada, Utah, Alaska.
8	Michigan, Ohio, West Virginia.
9	Wisconsin, Illinois, Indiana.
0	Colorado, Nebraska, North and South Dakota, Kansas, Minnesota, Iowa, Missouri.

EXAMINATION ADMINISTRATION

For all licenses *except the Novice operator*, the examinations are administered by three local amateur operators certified as volunteer examiners (VEs). Novice Class requires only two VEs, and they need not be certified. In fact, any two licensed General Class hams may give the Novice examination. VE teams usually provide the information on local VHF networks as to when and where examination sessions are to be held. Their efforts are coordinated by a VEC (volunteer-examiner coordinator) who accredits them to serve as a volunteer examiner. Nearly all VEs at testing sessions for Technician level and higher operator examinations will be glad to administer you the Novice test without charge. All you have to do is to appear at the testing session.

The test questions for all examination classes, including the Novice level, are developed and revised by the combined efforts of all VECs. The FCC no longer handles this function. Coordinating VECs make all questions and recommended multiple choice answers available to the public. They are widely published by various license preparation material publishers. All VECs use the exact same worded questions and there are no secret questions for any class of amateur operator license.

Except for the Novice Class, administering VEs may charge the candidate a test fee for certain reimbursable expenses incurred in preparing, processing and administering the amateur operator examination. The FCC annually adjusts the maximum amount of this test fee for inflation. After several adjustments, the fee is still only $5.25.

NO-CODE TECHNICIAN CLASS LICENSE EXAMINATION

As mentioned previously, the Technician Class new no-code examination will consist of a 2-part written examination. Most likely you will take both parts of the written examination in front of three accredited volunteer examiners in the same test session. Accredited volunteer examiners are registered through volunteer examiner coordinators to officially administer amateur operator examinations. Accredited volunteer examiners will hold an Advanced Class license or an Extra Class license. Accredited volunteers are uncompensated, and perform this testing service on a strictly volunteer basis. The VEs will help you fill out your FCC Form 610 (found in the back of this book) as described in Chapter 4. This way your application will be filed properly.

Your written examination will be taken from the Element 2 and Element 3A question pools in this book. There will be one examination with 30 questions taken from Element 2 and another examination with 25 questions taken from Element 3A. The actual breakdown of questions on your 30-question Element 2 exam, and on your 25-question Element 3A exam is shown in *Table 1-5*.

If you know the code and choose the Novice route, your Novice examination is taken in front of two General Class (or higher) ham buddies. They will make up their own written test, and their own code test. When you pass, they will both sign your FCC Form 610, and may mail it directly to the Federal Communications Commission for you. Thus, you will not have to appear at a scheduled VE test session, but set up your own schedule with your two General Class (or higher) buddies. If you took your Novice examination at a VE session and you plan to immediately upgrade by taking the Technician Plus written examination in front of three VEs, ask your Novice examiners for a photocopy of your FCC form 610 before they mail it to the FCC. The VEs must mail the form 610 to the FCC, but a photocopy of it will be accepted by the Technician Plus testing team.

For either route, all you need for the written examinations are the question pools in this book.

Technician Class Study Time

You should be able to prepare for the Technician Class written examinations in about four weeks. First study the Element 2 material, which one would study also if they were going the Novice route, and then study the Element 3A materials. When you take the 2-part examination, the VEs will first administer the Element 2 written examination, and then the Element 3A written examination. Your examiners will also offer an optional 5-wpm code test, and if you know the code, even a little bit—why not take the test, or at least give it a try!

The Question Pool

The Element 2 question pool contains 372 questions; the Element 3A question pool contains 326 questions. These questions, plus the multiple choice answers, one of which is a right answer, will be the precise, exact questions used on your upcoming examination. The volunteer examiners will not re-word the questions, nor will they change any of the wrong or right answers. This is good news! No secret questions, no strange answers, and absolutely no surprises once you have covered this book. Letter for letter, word for word, and number for number— the precise questions and answers contained in this book are those to be found on your upcoming examinations.

Both the Element 2 and Element 3A question pools has been recently updated by the Question Pool Committee (QPC), under the direction of Ray Adams, N4BAQ. The Question Pool Committee, and Adams, deserve tremendous credit in constantly keeping the question pool updated to agree with the latest FCC rules, and to match the current acceleration in amateur service technology with what is required to know for the examinations.

Chapter 3 is where you will find the two question pools. The Element 2 question pool is presented first, followed by the Element 3A question pool. When you take your Technician Class written examinations, the Element 2 exam will be administered first, followed by the Element 3A exam. If you know these two question pools, you will pass the examinations with flying colors!

The Actual Examination

The two written examinations you will take for your Technician class operator's license will consist of a total of 55 questions, and these questions are divided between topic categories as shown in *Table 1-5*. The Element 2 examination will consist of 30 questions taken from the Element 2 372-question pool, each with its multiple-choice answers. The Element 3A examination will consist of 25 questions taken from the Element 3A 326 question pool each with its multiple-choice answers. You must get at least 74% of the questions correct—22 of the 30 Element 2 questions, and 19 of the 25 Element 3A questions.

Table 1-5. Question Distribution for the Technician Class Exam

Topic		Number of Questions		
	Element 2		Element 3A	
Commission's Rules (FCC rules for the Amateur Radio services)	2A	10	3AA	5
Operating Procedures (Amateur station operating procedures)	2B	2	3AB	3
Radio Wave Propagation (Radio wave propagation characteristics)	2C	1	3AC	3
Amateur Radio Practices (Amateur Radio practices)	2D	4	3AD	4
Electrical Principles (Electrical principles as applied to amateur)	2E	4	3AE	2
Circuit Components (Amateur station equipment circuit components)	2F	2	3AF	2
Practical Circuits (Practical circuits employed in amateur station equipment)	2G	2	3AG	1
Signals and Emissions (Signals and emissions transmitted by amateur stations)	2H	2	3AH	2
Antennas and Feed Lines (Amateur station antennas and feed lines)	2I	3	3AI	3
TOTAL		30		25

Titles in parenthesis are the official subelement titles listed in FCC Part 97.

It's Easy!

Probably the primary prerequisite for passing any amateur operator license examination is the will to do it. If you follow the suggestions in this book, your chances are excellent. If you are fascinated by radio communication, and didn't want to learn the code, now is the time to join the amateur service as a Technician Class operator.

Technician Class License Privileges

ABOUT THIS CHAPTER

There is plenty of excitement out there on the amateur service bands with your Technician Class new no-code license, or your Technician Plus license. Just wait until you read about all of the band privileges you are going to get!

TECHNICIAN CLASS LICENSE

The Technician Class new no-code license is now the most popular way to get started in the amateur service. By taking two simple written examinations and scoring at least 74% on each exam, you will pass and be awarded a Technician Class operator's license. You will not need to take any type of code test to obtain your Technician Class license.

The Technician Class license privileges begin at 50 MHz in frequency and go higher from there. As shown in *Table 2-1*, you have full operating privileges on the 6-meter worldwide band; the 2-meter band, which is the world's most popular repeater band; the 222-MHz band where repeaters are linked together; the 440-MHz band where you can operate amateur television, satellite, and remote-base stations; and the 1270-MHz band with more amateur television, satellite communications, and repeater linking. On top of this, there are several other microwave bands on which you can experiment—all this, without a code test!

TECHNICIAN PLUS LICENSE

Remember, there are two paths to the Technician Plus operator license. If you are a Novice, all you need do is pass an Element 3A 25-question written examination at a VE test session to have Technician Plus privileges.

If you enter the amateur service as a Technician Class, all you need do is pass a 5-wpm Morse code (Element 1A) test at a VE test session, and you have Technician Plus privileges.

The Technician Plus operator has privileges on all frequency bands at and above 50 MHz, just like the Technician Class operator. But the Technician Plus operator also receives 10-meter worldwide voice and digital privileges, 10-meter CW privileges, 15-meter CW privileges, 40-meter CW privileges, and 80-meter CW privileges. 10 meters,

Table 2-1. Technician Class Operating Privileges

Wavelength Band	Frequency	Emissions	Comments
160 Meters	1800–2000 kHz	None	No privileges
80 Meters	3675–3725 kHz	None	No privileges
40 Meters	7100–7150 kHz	None	No privileges
30 Meters	10100–10150 kHz	None	No privileges
20 Meters	14000–14350 kHz	None	No privileges
15 Meters	21100–21200 kHz	None	No privileges
12 Meters	24890–24990 kHz	None	No privileges
10 Meters	28100–28500 kHz 28300–28500 kHz	None None	No privileges No privileges
6 Meters	50.0–54.0 MHz	All modes	Sideband voice, radio control, FM/FM repeater, digital computer, remote bases, and autopatches. Even CW. (1500 watts PEP output limitation)
2 Meters	144–148 MHz	All modes	All types of operation including satellite and owning repeater and remote bases. (1500 watt PEP output limitation)
1¼ Meters	222–225 MHz	All modes	All band privileges. (1500 watt PEP output limitation)
70 cm	420–450 MHz	All modes	All band privileges, including amateur television, packet, RTTY, FAX, and FM voice repeaters. (1500 watt PEP output.)
35 cm	902–928 MHz	All modes	All band privileges. Plenty of room! (1500 watt PEP output.)
23 cm	1240–1300 MHz	All modes	All band privileges.

15 meters, 40 meters, and 80 meters are those worldwide bands that let you communicate anywhere around the world, anytime, day or night, with a worldwide set.

TECHNICIAN OPERATING PRIVILEGES

But don't be disappointed if you put off the code test—the Technician Class operator has plenty of worldwide excitement on 6 meters, the first band we'll explore in detail.

6-METER WAVELENGTH BAND, 50.0-54.0 MHz

The Technician Class new no-code operator will enjoy all amateur service privileges and maximum output power of 1,500 watts on this worldwide band. Are you into radio control (R/C), and want to escape the interference on 74 MHz? On 6 meters, your Technician Class license allows you to operate on exclusive radio control channels at 50 MHz and 53 MHz, just for licensed hams.

On 6 meters, the Technician Class operator can get a real taste of long-range skywave skip communications. During the summer months, and during selected days and weeks out of the year, 50-MHz-54-MHz,

6-meter signals are refracted by the ionosphere, giving you incredible long-range communication excitement. During the summer months, it's almost a daily phenomena for 6 meters to skip all over the country. This is the big band for the Technician Class operator with this type of ionospheric, long-range, skip excitement. There's even repeaters on 6 meters. So make 6 meters "a must" at your future operating station.

Table 2-2 shows the ARRL 6-meter wavelength band plan:

Table 2-2. ARRL* 6-Meter Wavelength Band Plan, 50.0-54.0 MHz

MHz	Use
50.000–50.100	CW and beacons
50.060–50.080	Automatically controlled beacons
50.100–50.600	SSB
50.110	SSB DX calling frequency
50.200	SSB domestic calling frequency (Note: Suggest QSY up for local and down for long-distance QSOs)
50.600–51.000	Experimental and special modes
50.700	RTTY calling frequency
50.800–50.980	Radio Control (R/C) channels, 10 channels spaced 20 kHz apart (new)
51.000–51.100	Pacific DX window
51.000–52.000	Newly authorized FM repeater allocation.
51.100–52.000	FM simplex
52.000–52.050	Pacific DX window
52.000–53.000	FM repeater and simplex
53.000–54.000	Present radio control (R/C) channels, 10 channels spaced 100 kHz apart.

*Amateur Radio Relay League

2-METER WAVELENGTH BAND, 144 MHz-148 MHz

The 2-meter band is the world's most popular spot for staying in touch through repeaters. Here is where most all of those handheld transceivers operate, and the Technician Class operator receives unlimited 2-meter privileges!

Handie-talkie channels	Simplex autopatch	Rag-chewing
Transmitter hunts	Contests	Radio teleprinter
Autopatch	Traffic handling	Radio facsimile
Moon bounce	Satellite downlink	Emergency nets
Meteor bursts	Satellite uplink	Sporadic-E DX
Packet radio	Remote base	Aurora
Tropo-DX-ducting	Simplex operation	

The United States, and many parts of the world, are blanketed with clear, 2-meter, repeater coverage. They say there is nowhere in the United States you couldn't reach at least 1 or 2 repeaters with a little handheld transceiver. 2-meters has you covered!

The Technician Class license allows 1,500 watts maximum power output for specialized 2-meter communications, and also permits you to own and control a 2-meter repeater, too. *Table 2-3* gives the 2-meter wavelength band plan proposed by the ARRL VHF/UHF advisory committee:

Table 2-3. ARRL 2-Meter Wavelength Band Plan, 144-148 MHz

MHz	Use
144.00–144.05	EME (CW)
144.05–144.06	Propagation beacons (old band plan)
144.06–144.10	General CW and weak signals
144.10–144.20	EME and weak-signal SSB
144.200	National SSB calling frequency
144.20–144.30	General SSB operation, upper sideband
144.275–144.300	New beacon band
144.30–144.50	New OSCAR subband plus simplex
144.50–144.60	Linear translator inputs
144.60–144.90	FM repeater inputs
144.90–145.10	Weak signal and FM simplex
145.10–145.20	Linear translator outputs plus packet
145.20–145.50	FM repeater outputs
145.50–145.80	Miscellaneous and experimental modes
145.80–146.00	OSCAR subband—satellite use only!
146.01–147.37	Repeater inputs
146.40–146.58	Simplex
146.61–146.97	Repeater outputs
147.00–147.39	Repeater outputs
147.42–147.57	Simplex
147.60–147.99	Repeater inputs

Repeater frequency pairs (input/output):

144.61/145.21	144.89/145.49	146.40 or 146.60/147.00*
144.63/145.23	146.01/145.61	146.43 or 146.63/147.03*
144.65/145.25	146.04/146.64	146.46 or 146.66/147.06*
144.67/145.27	146.07/146.67	147.69/147.09
144.69/145.29	146.10/146.70	147.72/147.12
144.71/145.31	146.13/146.73	147.75/147.15
144.73/145.33	146.16/146.76	147.78/147.18
144.75/145.35	146.19/146.79	147.81/147.21
144.77/145.37	146.22/146.82	147.84/147.24
144.79/145.39	146.25/146.85	147.87/147.27
144.81/145.41	146.28/146.88	147.90/147.30
144.83/145.43	146.31/146.91	147.93/147.33
144.85/145.45	146.34/146.94	147.96/147.36
144.87/145.47	146.37/146.97	147.99/147.39

Additional channels available in large cities when 15 kHz and 20 kHz "splinter channels," interspersed between regular channels, are used.

*local option

1¼-METER WAVELENGTH BAND, 222 MHz-225 MHz

This popular band, shown in *Table 2-4,* is filled with activity with Novice Voice Class operators transmitting from 222.1 MHz to 223.91 MHz. The new Technician Class license permits you the use of the entire band, and 1,500 watts maximum output power instead of the 25 watts for Novice licensees. If you need some relief from the activity on 2 meters, the 222-225 MHz band is similar in propagation and use.

Table 2-4. ARRL 1¼-Meter Wavelength Band Plan, 222-225 MHz

MHz	Use	
220.00–220.05	EME (Earth-Moon-Earth)	
220.05–220.06	Propagation beacons	
220.06–220.10	Weak signal CW	
220.10	Calling frequency	ALLOCATED TO LAND MOBILE COMMERCIAL RADIO SERVICE
220.10–220.50	General weak signal, rag chewing and experimental communications	
220.50–221.90	Experimental and control links	
221.90–222.00	Weak-signal guard band	
222.00–222.10	CW only	
222.10–222.20	SSB, CW	
222.20	Calling frequency	
222.20–222.30	SSB, AM	
222.30–222.33	Beacons	
222.33–222.40	SSB, AM, PM, data (6 kHz bandwidth)	
222.40–223.40	FM repeater inputs	
223.40–223.50	FM simplex	
223.50	FM simplex calling frequency	
223.50–223.70	Packet (AFSK FM)	
223.70–224.00	High-speed digital (100 kHz per channel)	
224.00–225.00	FM repeater outputs	

Current 1.6-MHz separation retained.

Simplex frequencies (MHz):

223.42	223.52	223.62	223.72	223.82
223.44	223.54	223.64	223.74	223.84
223.46	223.56	223.66	223.76	223.86
223.48	223.58	223.68	223.78	223.88
223.50*	223.60	223.70	223.80	223.90

*National simplex frequency

Repeater frequency pairs (input/output)(MHz):

222.32/223.92	222.54/224.14	222.76/224.36	222.98/224.58	223.20/224.80
222.34/223.94	222.56/224.16	222.78/224.38	223.00/224.60	223.22/224.82
222.36/223.96	222.58/224.18	222.80/224.40	223.02/224.62	223.24/224.84
222.38/223.98	222.60/224.20	222.82/224.42	223.04/224.64	223.26/224.86
222.40/224.00	222.62/224.22	222.84/224.44	223.06/224.66	223.28/224.88
222.42/224.02	222.64/224.24	222.86/224.46	223.08/224.68	223.30/224.90
222.44/224.04	222.66/224.26	222.88/224.48	223.10/224.70	223.32/224.92
222.46/224.06	222.68/224.28	222.90/224.50	223.12/224.72	223.34/224.94
222.48/224.08	222.70/224.30	222.92/224.52	223.14/224.74	223.36/224.96
222.50/224.10	222.72/224.32	222.94/224.54	223.16/224.76	223.38/224.98
222.52/224.12	222.74/224.34	222.96/224.56	223.18/224.78	

The Federal Communications Commission is presently reallocating the bottom 2 MHz of this band, 220-222 MHz, to the land mobile commercial radio service. This reallocation occurred because the FCC did not feel there was enough amateur activity on the bottom 2 MHz. It is hoped that the Technician Class new no-code license will fill all the VHF and UHF bands with activity so we don't ever lose more frequencies due to inactivity!

70-cm WAVELENGTH BAND, 420-450 MHz

As you gain more experience on the VHF and UHF bands, you will soon be invited to the upper echelon of specialty clubs and organizations. The 450-MHz band is where the experts hang out. Amateur television (ATV) is very popular, so there's no telling who you may *see* as well as hear. This band also has the frequencies for controlling repeater stations and base stations on other bands, plus satellite activity. With a Technician Class license, you may even be able to operate on General Class worldwide frequencies if a General Class or higher control operator is on duty at the base control point. You would be able to talk on your 450-MHz handheld transceiver and end up in the DX portion of the 20-meter band. As long as the control operator is on duty at the control point, your operation on General Class frequencies is completely legal!

The 450 MHz band is also full of packet communications, RTTY, FAX, and all those fascinating FM voice repeaters. If you are heavy into electronics, you'll hear fascinating topics discussed and digitized on the 450-MHz band. A Technician Class operator has full power privileges as well as unrestricted emission privileges. *Table 2-5* presents the ARRL 70-cm (centimeter) wavelength band plan.

35-cm WAVELENGTH BAND, 902-928 MHz

Radio equipment manufacturers are just beginning to market equipment for this band. Many hams are already on the air using homebrew equipment for a variety of activities. If you are looking for a band with the ultimate in elbow room, this is it!

Table 2-6 shows the 35-cm wavelength band plan adopted by the ARRL Board of Directors in October, 1984.

Table 2-5. ARRL 70-cm Wavelength Band Plan, 420-450 MHz

MHz	Use
420.00–426.00	ATV repeater or simplex with 421.25-MHz video carrier control links and experimental
426.00–432.00	ATV simplex with 427.250-MHz video carrier frequency
432.00–432.07	EME (Earth-Moon-Earth)
432.07-432.08	Propagation beacons (old band plan)
432.08–432.10	Weak-signal CW
432.10	70-cm calling frequency
432.10–433.00	Mixed-mode and weak-signal work
432.30-432.40	New beacon band
433.00–435.00	Auxiliary/repeater links
435.00–438.00	Satellite only (internationally)
438.00–444.00	ATV repeater input with 439.250-MHz video carrier frequency and repeater links
442.00–445.00	Repeater inputs and outputs (local option)
445.00–447.00	Shared by auxiliary and control links, repeaters and simplex (local option); (446.0-MHz national simplex frequency)
447.00–450.00	Repeater inputs and outputs

Repeater frequency pairs (input/output is local option)(MHz):

442.000/447.000	442.600/447.600	443.200/448.200	443.800/448.800	444.400/449.400
442.025/447.025	442.625/447.625	443.225/448.225	443.825/448.825	444.425/449.425
442.050/447.050	442.650/447.650	443.250/448.250	443.850/448.850	444.450/449.450
442.075/447.075	442.675/447.675	443.275/448.275	443.875/448.875	444.475/449.475
442.100/447.100	442.700/447.700	443.300/448.300	443.900/448.900	444.500/449.500
442.125/447.125	442.725/447.725	443.325/448.325	443.925/448.925	444.525/449.525
442.150/447.150	442.750/447.750	443.350/448.350	443.950/448.950	444.550/449.550
442.175/447.175	442.775/447.775	443.375/448.375	443.975/448.975	444.575/449.575
442.200/447.200	442.800/447.800	443.400/448.400	444.000/449.000	444.600/449.600
442.225/447.225	442.825/447.825	443.425/448.425	444.025/449.025	444.625/449.625
442.250/447.250	442.850/447.850	443.450/448.450	444.050/449.050	444.650/449.650
442.275/447.275	442.875/447/875	443.475/448.475	444.075/449.075	444.675/449.675
442.300/447.300	442.900/447.900	443.500/448.500	444.100/449.100	444.700/449.700
442.325/447.325	442.925/447.925	443.525/448.525	444.125/449.125	444.725/449.725
442.350/447.350	442.950/447.950	443.550/448.550	444.150/449.150	444.750/449.750
442.375/447.375	442.975/447.975	443.575/448.575	444.175/449.175	444.775/449.775
442.400/447.400	443.000/448.000	443.600/448.600	444.200/449.200	444.800/449.800
442.425/447.425	443.025/448.025	443.625/448.625	444.225/449.225	444.825/449.825
442.450/447.450	443.050/448.050	443.650/448.650	444.250/449.250	444.850/449.850
442.475/447.475	443.075/448.075	443.675/448.675	444.275/449.275	444.875/449.875
442.500/447.500	443.100/448.100	443.700/448.700	444.300/449.300	444.900/449.900
442.525/447.525	443.125/448.125	443.725/448.725	444.325/449.325	444.925/449.925
442.550/447.550	443.150/448.150	443.750/448.750	444.350/449.350	444.950/449.950
442.575/447.575	443.175/448.175	443.775/448.775	444.375/449.375	444.975/449.975

Table 2-6. ARRL 35-cm Wavelength Band Plan, 902-928 MHz

MHz	Use
902–904	Narrow-bandwidth, weak-signal communications
902.0–902.8	SSTV, FAX, ACSB, experimental
902.8–903.0	Reserved for EME, CW expansion
903.0–903.05	EME exclusive
903.07–903.08	CW beacons
903.1	CW, SSB calling frequency
903.4–903.6	Crossband linear translator inputs
903.6–903.8	Crossband linear translator outputs
903.8–904.0	Experimental beacons exclusive
904–906	Digital communications
906–907	Narrow-bandwidth FM-simplex services, 25-kHz channels
906.5	National simplex frequency
907–910	FM repeater inputs paired with 919-922 MHz; 119 pairs every 25 kHz; e.g., 907.025, 907.050, 907.075, etc. 908-920 MHz uncoordinated pair.
910–916	ATV
916–918	Digital communications
918–919	Narrow-bandwidth, FM control links and remote bases
919–922	FM repeater outputs, paired with 907-910 MHz
922–928	Wide-bandwidth experimental, simplex ATV, Spread Spectrum

23-cm WAVELENGTH BAND, 1240-1300 MHz

There is plenty of over-the-counter radio equipment for this band, thanks to Novice Class operators getting onto those frequencies and exploring how far these microwaves go. Although the Novice operator is only allowed 5 watts, the Technician Class operator may run any amount of power—with 20 watts about the maximum limit.

The frequencies are in the microwave region, and this band is excellent to use with local repeaters in major cities.

Like the 450-MHz band and the 2-meter band, this band is sliced into many specialized operating areas. You can work orbiting satellites, operate amateur television, or own your own repeater with your Technician Class new no-code license. *Table 2-7* presents the 23-cm wavelength band plan adopted by the ARRL Board of Directors in January, 1985.

Table 2-7. ARRL 23-cm Wavelength Band Plan, 1240-1300 MHz

MHz	Use
1240–1246	ATV #1
1246–1248	Narrow-bandwidth FM point-to-point links and digital, duplex with 1258-1260 MHz
1248–1252	Digital communications
1252–1258	ATV #2
1258–1260	Narrow-bandwidth FM point-to-point links and digital, duplexed with 1246-1252 MHz
1260–1270	Satellite uplinks, reference WARC '79
1260–1270	Wide-bandwidth experimental, simplex ATV
1270–1276	Repeater inputs, FM and linear, paired with 1282-1288 MHz, 239 pairs every 25 kHz, e.g., 1270.025, 1270.050, 1270.075, etc. 1271.0-1283.0 MHz uncoordinated test pair
1276–1282	ATV #3
1282–1288	Repeater outputs, paired with 1270-1276 MHz
1288–1294	Wide-bandwidth experimental, simplex ATV
1294–1295	Narrow-bandwidth FM simplex services, 25-kHz channels
1294.5	National FM simplex calling frequency
1295–1297	Narrow bandwidth weak-signal communications (no FM)
1295.0–1295.8	SSTV, FAX, ACSB, experimental
1295.8–1296.0	Reserved for EME, CW expansion
1296.0–1296.05	EME exclusive
1296.07–1296.08	CW beacons
1296.1	CW, SSB calling frequency
1296.4–1296.6	Crossband linear translator input
1296.6–1296.8	Crossband linear translator output
1296.8–1297.0	Experimental beacons (exclusive)
1297–1300	Digital communications

10-GHz (10,000 MHz!) BANDS AND MORE

There are several manufacturers of ham microwave transceivers and converters for this range, so activity is excellent. Gunnplexers are the popular transmitter. Ten GHz is frequently used by hams to establish voice communications for controlling repeaters over paths from 20 miles to 100 miles using horn and dish antennas. Output power levels are usually less than one-eighth of a watt! It's really fascinating to see how directional the microwave signals are. If you live on a mountaintop, 10 GHz is for you.

All modes and licensees except Novices are authorized on the bands shown in *Table 2-8*. There is much Amateur Radio experimentation on these bands.

Table 2-8 Gigahertz Bands

2.30–2.31 GHz	10.0–10.50 GHz*	165.0–170.0 GHz
2.39–2.45 GHz	24.0–24.25 GHz	240.0–250.0 GHz
3.30–3.50 GHz	48.0–50.00 GHz	All above 300 GHz
5.65–5.925 GHz	71.0–76.00 GHz	

*Pulse not permitted

TECHNICIAN PLUS OPERATING PRIVILEGES

If you once learned the code, or enjoy Morse code, think about taking the 5-wpm code test for Technician Plus added privileges. This chart highlights the added privileges you will receive on 80 meters, 40 meters, 15 meters, and 10 meters by passing a simple Morse code test.

Table 2-9. Technician Plus Class Operating Privileges

Wavelength Band	Frequency	Emissions	Comments
160 Meters	1800–2000 kHz	None	No privileges
80 Meters	3675–3725 kHz	Code only	**Limited to Morse code (200 watt PEP output limitation)
40 Meters	*7100–7150 kHz	Code only	**Limited to Morse code (200 watt PEP output limitation)
30 Meters	10100–10150 kHz	None	No privileges
20 Meters	14000–14350 kHz	None	No privileges
15 Meters	21100–21200 kHz	Code only	**Limited to Morse code (200 watt PEP output limitation)
12 Meters	24890–24990 kHz	None	No privileges
10 Meters	28100–28500 kHz 28100–28300 kHz 28300–28500 kHz	Code Data and code Phone and code	Morse code and digital computer (200 watt PEP output code limitation) Sideband voice (200 watt code PEP output limitation)
6 Meters	50.0–54.0 MHz	All modes	Morse code, sideband voice, radio control, FM/FM repeater, digital computer, remote bases, and autopatches (1500 watts PEP output limitation)
2 Meters	144–148 MHz	All modes	All types of operation including satellite and owning repeater and remote bases. (1500 watt PEP output limitation)
1¼ Meters	220–225 MHz	All modes	All band privileges. (1500 watt PEP output limitation)
70 cm	420–450 MHz	All modes	All band privileges, including amateur television, packet, RTTY, FAX, and FM voice repeaters. (1500 watt PEP output.)
35 cm	902–928 MHz	All modes	All band privileges. Plenty of room! (1500 watt PEP output.)
23 cm	1240–1300 MHz	All modes	All band privileges.

* U.S. licensed operators in other than our hemisphere (ITU Region 2) are authorized 7050-7075 kHz due to short wave broadcast interference.

** No change from Novice Privileges.

80-METER WAVELENGTH BAND, 3500-4000 kHz

Novice and Technician Plus Class privileges on the 80-meter band are Morse code only from 3675 to 3725 kHz. These frequencies have recently been shifted to accommodate our Canadian neighbor's band plan. It takes a code test from Technician Class to get onto this band for CW work.

40-METER WAVELENGTH BAND, 7000-7300 kHz

Passing a 5-wpm code test, gives you Morse code only privileges on this band at 7100 to 7150 kHz. This is a popular Novice and Technician Plus CW band because evening QSO's can reach up to 5,000 miles away.

15-METER WAVELENGTH BAND, 21,000-21,450 kHz

Technician Plus Class via the 5-wpm code test provides you CW privileges from 21,100 to 21,200 kHz in this portion of the worldwide band. Novice and Technician Plus operators can expect ranges to distant stations in excess of 10,000 miles during daylight hours. Nighttime DX activity on this band normally fades away about 9:00 P.M. local time.

10-METER WAVELENGTH BAND, 28,000-29,700 kHz

Passing a code test at 5 wpm allows you to operate on this very popular band for regular worldwide privileges. Novice and Technician Plus operators may operate code and digital computer privileges between 28,100 to 28,300 kHz, and monitor 28,200 to 28,300 kHz for low-power propagation beacons. Novice and Technician Plus operators, having passed a code test, are allowed code and voice privileges between 28,300 to 28,500 kHz. It's worth it to learn the code for this popular band after you have successfully passed your Technician Class license test.

ABOUT THE MORSE CODE TESTS

To obtain Technician Plus after you have passed the two written examinations for your Technician Class operator's license requires you to pass a 5-wpm Morse code test. You may be interested in what the 5-wpm code test is like, and what additional privileges the incentive licensing of the amateur service provides if you were to study Morse code and gain good proficiency in sending and receiving it.

5-wpm Morse Code Test

Part 97 details how Element 1A should be prepared. While the rules state that the code test should be able to prove the applicant's ability to transmit correctly by hand and to receive text by ear in the International Morse code at 5 wpm, the FCC allows a volunteer examiner to dispense with the sending portion of the code test. It has been their experience that if you can copy 5 wpm, then you can also send it with a hand key. It remains optional with the VE whether or not they want to require sending code by hand in the test. Most VEs don't.

Your Element 1A code test will be sent at a rate of 5 wpm. The actual dots and dashes will be generated at 15 wpm, but with big spaces between to give you a second or so to think of the letter just sent.

Test Content

Effective January 1, 1987, the optional code test *must* contain *all* alphabet letters, numerals Ø through 9, certain punctuation—the period, comma, question mark and slant bar, plus the operating procedure signs $\overline{AR}$, $\overline{SK}$, and $\overline{BT}$. Originally, the applicant was responsible for knowing them, but was not necessarily tested on all characters. No longer. *The test must contain all of them!*

It is perfectly legal for volunteer examiners to obtain a correctly designed prepackaged code examination transmission from other sources as long as the code message is unknown to the applicant and *the preparer is a General Class amateur operator or higher.*

The volunteer examiner team (remember, it takes *two*) is given wide latitude in constructing the telegraphy examination. It will most likely take the form of a typical over-the-air code transmission made by amateurs. This is called a "QSO-type" code examination. Amateurs normally swap information concerning their call signs, names, locations (QTH), equipment used, weather, RST signal report, occupations, and the like, when communicating in CW. VEs are prohibited from using well known phrases in their code examinations for obvious reasons.

Test Time and Format

The Element 1A optional code test transmission will probably run for about five to eight minutes. While it is *customary* to pass an applicant if he/she can copy 25 characters in a row, or answer seven out of ten questions correctly about the transmission, neither one is *specifically stipulated* in the rules. Morse code tests with multiple choice, true-false, or fill-in-the-blank questions also are acceptable. In any event, it is the *VE team* that decides on the format of the Morse code test.

It is perfectly acceptable to take (or administer) amateur operator code tests at computer keyboards. To aid the examiner, there are some excellent personal computer testing software programs in the marketplace. Most personal computers not only have the capability to generate Morse code at a prescribed speed, but code tests of various formats as well.

The rules state in Part 97 that five characters shall constitute a word. If there are punctuation and numerals (but, strangely, not operating procedure signs), they are counted as two characters; therefore, 25 characters sent at 5-wpm passes. The VEs need only be satisfied that you can *understand* what is being transmitted. The important point is—does the applicant know what is being sent? *It is the sole responsibility of the VEs to make this determination.*

Learning Morse Code

Here are three suggested ways for you to learn Morse code:

1. Use the cassette tapes in the *New Novice Voice Class* license preparation materials available in the Radio Shack stores.
2. As you operate with your Technician Class license, tune in to some live code broadcasts on your bands and listen to typical code transmissions.
3. Use a code key and oscillator for practice by yourself and with a friend.

The *New Novice Voice Class* license preparation materials referred to above has two cassettes and a book containing the Element 2 question pool. The tapes are specifically design to bring a person's proficiency from zero to sending and receiving Morse code at least at 5 wpm. *Table 2-10* shows what the tapes contain.

Table 2-10. Sequence of Lessons on Cassettes

■ Lesson 1	E T M A N I S O $\overline{\text{SK}}$ Period
■ Lesson 2	R U D C 5 Ø $\overline{\text{AR}}$ Question Mark
■ Lesson 3	K P B G W F H $\overline{\text{BT}}$ Comma
■ Lesson 4	Q L Y J X V Z $\overline{\text{DN}}$ 1 2 3 4 6 7 8 9
■ Lesson 5	Random code with narrated answers
■ Lesson 6	A typical Novice code test

Working with cassette tapes such as these for just 10 to 20 minutes per day, you can easily accomplish sending and receiving code at 5 wpm.

Going Beyond 5 WPM

After you have successfully received, and possibly sent, the Morse code at 5 wpm at a VE test session, you will be passed on your Element 1A code requirement and allowed to operate as a Technician Plus Class operator.

If your interest in CW sending and receiving has been sparked by your success with 5 wpm, you might consider progressing to the next step—as a General Class operator. The General Class code test is administered the same way as the 5-wpm code test except it must meet the requirements of Element 1B. For Element 1B (General), the code speed is 13 wpm, with the actual characters sent at 15 wpm. No sending test is required by the FCC, although your examiners have the option of requiring it.

Passing the 13-wpm code test allows you voice privileges on all of the worldwide bands with a General Class license. If you plan to cruise on a boat throughout the world, the General Class ticket is just for you. If you want to work the world, day and night on worldwide frequencies, you ultimately want to pass the 13-wpm code test. Of course, there is a 25-question written examination on Element 3B that also must be passed before you can operate as a General Class licensee. There are license preparation materials for General Class—*New General Class*—available at Radio Shack with two cassettes and a book containing the Element 3B question pool to aid in passing the General Class code test and written examination.

There is another step after 13 wpm. It is necessary for an Extra Class operator's license. For the Extra Class license, the very highest ham radio license available, the code requirement is 20 wpm, sent at 21-wpm character rate. Just like all other exams, one minute of perfect copy or seven out of ten questions answered correctly passes the code test to Element 1C requirements. Passing the 20-wpm code test opens up every single frequency available to any class of amateur operator. Two written examinations must be passed to obtain an Extra Class license, after you have met the qualifications for a General Class license. A 50-question Element 4A written examination for the Advanced Class operator license, and then an Extra Class 40-question Element 4B examination. When you have passed the extra code and written tests, you have achieved the highest grade amateur operator's license there is.

SUMMARY

Enjoy the excitement of being a ham radio operator and communicate worldwide with other amateur operators without a code test. Your Technician Class operator license allows you all ham operator privileges on all bands with frequencies greater than 50 MHz. You have operating privileges on the worldwide 6-meter band, on the world's popular 2-meter and 222-MHz repeater bands, on the amateur television, satellite communications, and repeater linking 440-MHz and 1270-MHz bands, and on the line-of-sight microwave bands at 10 GHz and above.

And if you know some Morse code, you can pass an Element 1A 5-wpm code test and become a Technician Plus operator with voice privileges on 10 meters, and long-range CW privileges on 10 meters, 15 meters, 40 meters and 80 meters.

Getting Ready for the Examination

ABOUT THIS CHAPTER

Your examinations for the Technician Class operator will be taken from the pool of 372 Element 2 questions, with multiple-choice answers, and 326 Element 3A questions, with multiple-choice answers.

The written Element 2 examination will contain exactly 30 questions. You must get 74% of the answers correct. That means that you must answer 22 questions correctly, or miss no more than 8 questions.

On the Element 3A examination, 25 questions will be selected word for word from the 326-question pool. You must score at least 74%, which means you must answer 19 questions correctly, or miss no more than 6 questions.

These two examinations may be administered one after another, in ascending order. You must pass Element 2 before you are administered Element 3A. If you should fail the Element 2 examination, your testing team may allow you to try another examination immediately. You might arrange to take the two examinations on separate days based on the VEs' schedule.

The questions and answers will be identical to what will follow in this chapter. Each question will be word for word, with four possible word-for-word, multiple-choice answers, one of which will be correct. The only thing you won't find on the test is the listed correct answer and explanations that you find here in the book!

WHAT THE EXAMINATIONS CONTAIN

The examination questions of all licensed class levels and their appropriate multiple-choice distractors are public information. They are widely published, and are identical to what is in this book. FCC rules prohibit any examiner or examination team from making *any changes* to any questions, including any numerical values. No numbers, words, letters, or punctuation marks can be altered from the published question pool. By studying the Element 2 and Element 3A question pools in this book, you will be reading the same exact questions that will appear on your 30-question Element 2 written examination, and your 25-question Element 3A examination.

Table 3-1 shows how the Element 2 30-question examination and Element 3A 25-question examination will be selected from the two question pools. Each pool is divided up into various subelements. Each subelement covers a different subject. For example, for the Element 2 examination, 10 questions out of 30 will be taken from the Commission's Rules subelement. For the Element 3A examination, 5 questions out of 25 will be taken from the Commission's Rules subelement.

All volunteer examination teams use the same pool of questions and multiple-choice answers that are in this chapter. This uniformity in study material insures common examinations throughout the country. Most examinations are computer-generated, and the computer selects the right number of questions out of each subelement for your two upcoming examinations.

Table 3-1. FCC Question Pool

Element 2 Subelement		Page	Total Questions	Examination Questions
2A	Commission's Rules	32	114	10
2B	Operating Procedures	59	48	2
2C	Radio Wave Propagation	70	18	1
2D	Amateur Radio Practices	74	45	4
2E	Electrical Principles	85	44	4
2F	Circuit Components	94	21	2
2G	Practical Circuits	99	20	2
2H	Signals and Emissions	104	23	2
2I	Antennas and Feed Lines	110	39	3
	TOTALS		372	30

Element 3A Subelement		Page	Total Questions	Examination Questions
3AA	Commission's Rules	121	64	5
3AB	Operating Procedures	137	30	3
3AC	Radio Wave Propagation	145	30	3
3AD	Amateur Radio Practices	153	44	4
3AE	Electrical Principles	163	34	2
3AF	Circuit Components	172	37	2
3AG	Practical Circuits	181	17	1
3AH	Signals and Emissions	186	28	2
3AI	Antennas and Feed Lines	191	42	3
	TOTALS		326	25

All amateur service examination questions are reviewed periodically by a team of volunteer examination coordinators assembled as a Question Pool Committee (QPC). These are fellow hams with Extra Class licenses who volunteer their time to insure today's tests are accurate and fair, and are subject matter relevant. As technology changes, so will the questions. A public notice will be given approximately one year in advance of any question changes so book publishers can revise their study materials.

> THIS BOOK CONTAINS ALL REVISIONS FOR THE
> TECHNICIAN CLASS NEW NO-CODE LICENSE
> Effective date: February 14, 1991.

QUESTION CODING

Each question of the 372 Element 2 question pool and the 326 Element 3A question pool is coded according to the numbers and letters assigned to each precise question. *Figure 3-1* explains the question coding using the Element 3A as an example.

The coded numbers and letters reveal important facts about each question. Each subelement has a number of topics within it. Each topic is numbered, and the questions in each topic are numbered. Therefore, in *Figure 3-1a*, 3AB-1.2 indicates the question is an Element 3A question from subelement B (Operating Procedures). It is the second (.2) question for topic number 1 (RST reporting system).

If the question has a subtopic, as shown in *Figure 3-1b*, the subtopics are numbered and the questions in each subtopic are numbered. Question 3AB-6-1.1 is from the same subelement 3AB, but is the first question for subtopic number 1 in topic number 6.

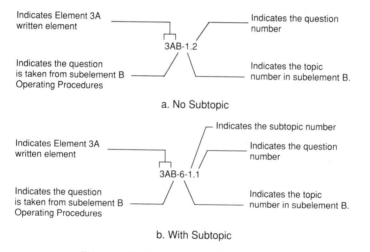

Figure 3-1. Examination Question Coding

NOTE: QUESTION POOL CHANGES

The question pools have been revised to better agree with the Part 97 FCC rules. The FCC has streamlined the language within the rules which dramatically makes studying for the Technician Class license examination easier! No longer will you need to memorize letter-number-letter combinations like "F1B," or "J3E." All examinations will now refer to these letter-number-letter designators in plain language, such as "digital communications," and "single-sideband voice communications." We congratulate the question pool committee (QPC), under the direction of Ray Adams, N4BAQ, for their continued efforts to constantly keep the FCC question pool updated and current with the latest changes of technology.

A QUICK LOOK AT EXAM QUESTIONS

Go ahead now and scan through the Element 2 and Element 3A questions. As you scan through, note the subelements we talked about. Note also that the number of examination questions to be taken from the subelement is given. This corresponds to the number in *Table 3-1* for the subelement. Notice that the question coding, which we explained, indicates when subelements and topics change.

Be aware that there is only one correct answer. Notice also that the correct answer and the correct answer explanation are given right after the question so that you have an easy time learning the material.

HINTS ON STUDYING FOR THE EXAM

All of the material that you need to study for your Technician Class examinations is in this chapter, organized and combined in a convenient format. Here are some suggestions to make your learning easier:

1. Begin with Subelement A of each of the Element 2 and Element 3A question pools. Study the question, the correct answer, and the brief explanation that we have given to better understand the importance of the question and why the answer is correct. It's also a good idea to look at the incorrect answers for some additional pointers on why the correct answer is indeed the right one.

   Element 2 begins with subelement 2A—Commission's Rules— and Element 3A begins with subelement 3AA—Commission's Rules. There are 372 total questions in the Element 2 question pool, and 326 questions in the Element 3A question pool.

2. Since the Technician Class requires two written examinations, it may be best to first study Element 2, and then have someone give you a trial examination according to the distribution shown in *Table 3-1*. Once you have Element 2 down cold, you can move on to Element 3A to complete your study for both written examinations. You will be required to take the Element 2 examination before the Element 3A.

   If you think you might forget what you learned in Element 2, contact an accredited VEC examination team, or any other two ham buddies with a General or higher license, and arrange to take Novice Element 2 all by itself. You will have 365 days time to pass either the Element 3A written examination to become a Technician Class operator, or the Novice Element 1A code test to completely validate the Novice license.

   In the Gordon West Radio School classes, Element 2 is taught first, then an outside examination team comes in and tests the students just on Element 2. Next, the school concentrates on Element 3A. If the 3A written examination is passed within 10 days of the Element 2 examination, the examiners would sign the same 610 form, validating a Technician Class license pass.

3. Be sure to read over each multiple-choice answer carefully. Some start out looking good, but just one or two words may change the answer from right to wrong. Also, don't anticipate that the multiple-choice answers will always appear in the exact same A-B-C-D order on your actual Technician Class examinations.
4. Keep in mind how many questions may be taken out of any one subelement.
5. A fun way of preparing for the exam is to let someone else read the correct answer, and you try to recite the exact question! This works out particularly well as a group exercise.
6. Once you have most of the questions "down cold," begin checking them off your list. Put a check beside each question you know perfectly. Then concentrate on those harder questions that may require some memorization and better understanding. Check (✓) them off as you are sure you know them.
7. Try a practice examination and see how well you do. Maybe you can get some ham buddies to make one up for you.

EXAMINATION QUESTION POOL

Are you ready to prepare for your new Technician Class license? Here are the two question pools. Let's get started learning these questions and answers. Study hard, and good luck!

When you complete your study, head for the local accredited VE team to take your 2-part examination in one sitting. This is a good way to go, and gets everything out of the way all at once.

Subelement 2A – Commission's Rules (10 examination questions from 114 questions in 2A)

One question must be from the following 10 questions:

2A-1.1 What are the five principles that express the fundamental purpose for which the amateur service rules are designed?
A. Recognition of emergency communications, advancement of the radio art, improvement of communication and technical skills, increase in the number of trained radio operators and electronics experts, and the enhancement of international goodwill
B. Recognition of business communications, advancement of the radio art, improvement of communication and business skills, increase in the number of trained radio operators and electronics experts, and the enhancement of international goodwill
C. Recognition of emergency communications, preservation of the earliest radio techniques, improvement of communication and technical skills, maintain a pool of people familiar with early tube-type equipment, and the enhancement of international goodwill
D. Recognition of emergency communications, advancement of the radio art, improvement of communication and technical skills, increase in the number of trained radio operators and electronics experts, and the enhancement of a sense of patriotism

ANSWER A: Memorize these five principles! They best describe what the ham radio service is for you, for other hams, and for the world we serve. We are there for emergencies; we help develop new radio techniques; we're always coming up with new ways to communicate; we develop ourselves into good, trained radio operators; and our worldwide communications with fellow hams in other countries should always enhance international goodwill.

2A-1.2 Which of the following is *not* one of the basic principles for which the amateur service rules are designed?
A. Providing emergency communications
B. Improvement of communication and technical skills
C. Advancement of the radio art
D. Enhancement of a sense of patriotism

ANSWER D: This is a reverse question, so look for the WRONG answer! Although ham radio is a fun and public service hobby, it may not necessarily enhance "patriotism."

2A-1.3 The amateur service rules were designed to provide a radio communications service that meets five fundamental purposes. Which of the following is *not* one of those principles?
A. Improvement of communication and technical skills
B. Enhancement of international goodwill
C. Increase the number of trained radio operators and electronics experts
D. Preserving the history of radio communications

ANSWER D: Here's another backwards question where you need to find the wrong answer. No, ham radio does not necessarily preserve the history of radio communications. (Although you might think so if you could see all of the old equipment your author still has in service!)

2A-1.4 The amateur service rules were designed to provide a radio communications service that meets five fundamental purposes. What are those principles?

A. Recognition of business communications, advancement of the radio art, improvement of communication and business skills, increase in the number of trained radio operators and electronics experts, and the enhancement of international goodwill
B. Recognition of emergency communications, advancement of the radio art, improvement of communication and technical skills, increase in the number of trained radio operators and electronics experts, and the enhancement of international goodwill
C. Recognition of emergency communications, preservation of the earliest radio techniques, improvement of communication and technical skills, maintain a pool of people familiar with early tube-type equipment, and the enhancement of international goodwill
D. Recognition of emergency communications, advancement of the radio art, improvement of communication and technical skills, increase in the number of trained radio operators and electronics experts, and the enhancement of a sense of patriotism

ANSWER B: Memorize these principles. They truly sum up what ham radio is all about. Read over the incorrect answers, and see the hidden wording that makes them wrong.

2A-2.1 What is the definition of the amateur service?

A. A private radio service used for personal gain and public benefit
B. A public radio service used for public service communications
C. A radiocommunication service for the purpose of self-training, intercommunication and technical investigations
D. A private radio service intended for the furtherance of commercial radio interests

ANSWER C: The Federal Communications Commission rules, Part 97, describes the amateur service as a radiocommunications service for the purpose of *self-training, intercommunication and technical investigation*, carried out by amateurs, that is, duly authorized persons interested in radio technique solely with a personal aim and without pecuniary interest. While it is true that ham operators are known for their emergency communications assistance in times of need, the amateur service is basically a *hobby radio service*, just for fun!

2A-2.2 What name is given to the radiocommunication service that is designed for self-training, intercommunication, and technical investigation?

A. The amateur service
B. The Citizen's Radio Service
C. The Experimenter's Radio Service
D. The Maritime Radio Service

ANSWER A: We sometimes call the amateur service "ham radio." What makes our hobby fun is experimentation and training ourselves in front of our ham radios.

2A-3.1 What document contains the specific rules and regulations governing the amateur service in the United States?

A. Part 97 of Title 47 CFR (Code of Federal Regulations)
B. The communications act of 1934 (as amended)

C. The Radio Amateur's Handbook

D. The minutes of the International Telecommunication Union meetings

ANSWER A: A copy of Part 97 of Title 47 of the Code of Federal Regulations should be in every amateur's library.

2A-3.2 Which one of the following topics is *not* addressed in the rules and regulations of the amateur service?

A. Station operation standards

B. Technical standards

C. Providing emergency communications

D. Station construction standards

ANSWER D: Station construction standards are not addressed in the FCC rules and regulations; however, assemble your station carefully and safely, insuring that everything is well-grounded.

2A-4.1 What is the definition of an amateur operator?

A. A person who has not received any training in radio operations

B. A person holding a written authorization to be the control operator of an amateur station

C. A person who performs private radio communications for hire

D. A trainee in a commercial radio station

ANSWER B: This will be *you* when you pass your 30-question Novice elementary theory exam and a 5 words-per-minute Morse code exam before two volunteer ham examiners. You will hold a valid ten-year license to operate an amateur station. The license will be issued by the Federal Communications Commission.

2A-4.2 What term describes a person holding a written authorization to be the control operator of an amateur station?

A. A Citizen Radio operator

B. A Personal Radio operator

C. A Radio Service operator

D. An amateur operator

ANSWER D: We sometimes call ourselves "ham radio operators," which is the same thing as an amateur operator.

One question must be from the following 10 questions:

2A-5.1 What is the portion of an amateur operator/primary station license that conveys operator privileges?

A. The verification section

B. Form 610

C. The operator license

D. The station license

ANSWER C: You will receive your operator license and station call sign directly from the Federal Communications Commission. You will be the only one in the world with that call sign. It takes about 30 days after you pass your license test to receive your call letters. Make several copies of your license, and always keep a copy of your operator license with you when you operate your equipment.

2A-5.2 What authority is derived from an amateur operator/primary station license?

A. The authority to operate any shortwave radio station
B. The authority to be the control operator of an amateur station
C. The authority to have an amateur station at a particular location
D. The authority to transmit on either amateur or Class D citizen's band frequencies

ANSWER B: It takes about 30 days for your FCC call letters to be sent to you from the Federal Communications Commission in Gettysburg, Pennsylvania. It's a two-part license: operator license and station license.

2A-6.1 What authority is derived from a written authorization for an amateur station?

A. The authority to use specified operating frequencies
B. The authority to operate an amateur station
C. The authority to enforce FCC Rules when violations are noted on the part of other operators
D. The authority to transmit on either amateur or Class D citizen's band frequencies

ANSWER B: Your station license authorizes you to use any type of Amateur Radio equipment at any particular location. You may operate your radio equipment almost anywhere in the United States and its possessions, except in commercial airplanes. But *you don't have to operate* your station at the address you list on your application as your current primary station location.

2A-6.2 What part of your amateur license gives you authority to operate an amateur station?

A. The operator license
B. The FCC Form 610
C. The station license
D. An Amateur Radio license does not specify a station location.

ANSWER C: Always carry a copy of your amateur operator/primary station license with you when you operate your set. This allows you to operate from any particular location you want, including the exact location shown on the station license.

2A-7.1 What is an amateur station?

A. A licensed radio station engaged in broadcasting to the public in a limited and well-defined area
B. A radio station used to further commercial radio interests
C. A private radio service used for personal gain and public service
D. A station in an amateur service consisting of the apparatus necessary for carrying on radiocommunications

ANSWER D: Your ham station will consist of a radio device, at a particular location, that will be used for amateur communications. This station can go anywhere that you go, and might be mobile, a base, or a handheld set. You also are allowed to choose any authorized frequency in any authorized band. You can change radio equipment type at anytime.

2A-8.1 Who is a control operator?

A. An amateur operator designated by the licensee of a station to be responsible for the transmission from that station to assure compliance with the FCC rules
B. A person, either licensed or not, who controls the emissions of an amateur station

C. An unlicensed person who is speaking over an amateur station's microphone while a licensed person is present

D. A government official who comes to an amateur station to take control for test purposes

ANSWER A: This is a fancy name for you, the person who holds an amateur operator/primary station license. You, or any other ham you designate, are in control of all transmissions, and are responsible for the proper operation of the station.

2A-8.2 If you designate another amateur operator to be responsible for the transmissions from your station, what is the other operator called?
A. Auxiliary operator
B. Operations coordinator
C. Third-party
D. Control operator

ANSWER D: If you are designated a control operator, you have a big responsibility ahead of you. Always remember that you are responsible for the proper operation of the station no matter who talks over the mike.

2A-9.1 List the five United States amateur operator license classes in order of increasing privileges.
A. Novice, General, Technician, Advanced, Amateur Extra
B. Novice, Technician, General, Advanced, Digital
C. Novice, Technician, General, Amateur, Extra
D. Novice, Technician, General, Advanced, Amateur Extra

ANSWER D: Memorize the order of amateur operator licenses. Remember, all ham licenses are additive—you can't skip over any license category to get to the top one.

2A-9.2 Which US amateur operator license is considered to be the "entry level" or "beginner's" license?

This question has been retracted by the QPC because the FCC established, effective February 14, 1991, the Technician Class as a no-code license class.

This question is not to be used on any written examination administered after February 14, 1991 until or unless it is revised or replaced by the QPC.

2A-9.3 What is the license class immediately above Novice Class?
A. The Digital Class license
B. The Technician Class license
C. The General Class license
D. The Experimenter's Class license

ANSWER B: If you started with Novice, don't stop! Once you have passed the Novice examination, begin studying the Element 3A question pool that is just as much fun to read as this one. Your next step is Technician Plus. If you are going directly for Technician Class, no code test is required.

One question must be from the following 10 questions:

2A-10.1 What frequencies are available in the amateur 80-meter band for a control operator holding a Novice Class operator license?

This question has been retracted by the QPC because the FCC established, effective February 14, 1991, the Technician Class as a no-code license class.

This question is not to be used on any written examination administered after February 14, 1991 until or unless it is revised or replaced by the QPC.

2A-10.2 What frequencies are available in the amateur 40-meter band for a control operator holding a Novice Class operator license in ITU region 2?
A. 3500 to 4000 kHz
B. 3675 to 3725 kHz
C. 7100 to 7150 kHz
D. 7000 to 7300 kHz

ANSWER C: This is the 40-meter wavelength band, so memorize the frequencies. The 40-meter band is great at nighttime for long-range, Novice, CW contacts. During the day, CW contacts may be made up to 400 miles away.

2A-10.3 What frequencies are available in the amateur 15-meter band for a control operator holding a Novice class operator license?
A. 21.100 to 21.200 MHz
B. 21.000 to 21.450 MHz
C. 28.000 to 29.700 MHz
D. 28.100 to 28.200 MHz

ANSWER A: Here's where the CW fun begins for the Novice on 15 meters—during the day and evening. You can usually exceed 3,000 miles with a telegraph key!

2A-10.4 What frequencies are available in the amateur 10-meter band for a control operator holding a Novice Class operator license?
A. 28.000 to 29.700 MHz
B. 28.100 to 28.300 MHz
C. 28.100 to 28.500 MHz
D. 28.300 to 28.500 MHz

ANSWER C: 10 meters is a "daytime" band. It's divided in half. 28.3 to 28.5 MHz is where most Novices hang out for long-range voice contacts. 28.1 to 28.2 MHz is a great place for CW. 28.2 to 28.3 MHz is where automatic beacons transmit propagation CW signals.

2A-10.5 What frequencies are available in the amateur 220-Mhz band for a control operator holding a Novice Class operator license in ITU region 2?
A. 225.0 to 230.5 MHz
B. 222.1 to 223.91 MHz
C. 224.1 to 225.1 MHz
D. 222.2 to 224.0 MHz

ANSWER B: Plenty of repeater activity here. Memorize the 220-MHz band frequencies.

2A-10.6 What frequencies are available in the amateur 1270-Mhz band for a control operator holding a Novice Class operator license?
A. 1260 to 1270 MHz
B. 1240 to 1300 MHz
C. 1270 to 1295 MHz
D. 1240 to 1246 MHz

ANSWER C: This band is just becoming popular. Your author uses it for repeater operation throughout Southern California. It's an exceptional band for small handhelds and mobile units in the big city where signals become quite reflective. Memorize these frequencies.

2A-10.7 If you are operating your amateur station on 3725 kHz, in what meter band are you operating?
    A. 80 meters
    B. 40 meters
    C. 15 meters
    D. 10 meters

ANSWER A: Look at the frequency privileges chart in the appendix of this book, and memorize the MHz frequencies as they relate to the band in meters. To convert MHz to meters, first convert kHz (3725) to MHz (3.725). Now divide this into 300, and presto, you end up with 80.5 meters. The answer is rounded to 80 because the wavelength for a band is an average number broadly covering all the frequencies in the band. The meter band is the wavelength of the operating frequency. Wavelength is found by the equation:

$$\text{meters} = \frac{300}{f}$$

which says that the wavelength in meters is equal to 300 divided by the frequency in megahertz. With a calculator, the keystrokes are: CLEAR 300 ÷ 3.725 = . The answer is 80.5 meters.

$$\text{Wavelength (in meters)} = \frac{300}{f \, (MHz)}$$

$$\text{or } f \, (\text{in MHz}) = \frac{300}{\text{wavelength (in meters)}}$$

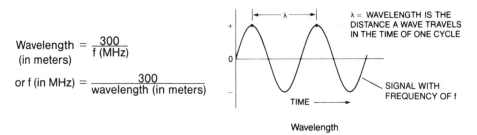

λ = WAVELENGTH IS THE DISTANCE A WAVE TRAVELS IN THE TIME OF ONE CYCLE

SIGNAL WITH FREQUENCY OF f

Wavelength

2A-10.8 If you are operating your amateur station on 7125 kHz, in what meter band are you operating?
    A. 80 meters
    B. 40 meters
    C. 15 meters
    D. 10 meters

ANSWER B: Novices in North and South America may operate their transmitters *using telegraphy only* between 7100 and 7150 kHz. The frequency 7125 kHz is located in the 40-meter ham band. The equation used for calculating is shown in 2A-10.7. The calculator keystrokes are: CLEAR 300 ÷ 7.125 = . The answer, 42.1, is rounded to 40 meters.

2A-10.9 If you are operating your amateur station on 21150 kHz, in what meter band are you operating?
    A. 80 meters
    B. 40 meters

C. 15 meters
D. 10 meters

ANSWER C: Convert 21,150 kHz into MHz. That's right, 21.150 MHz. Now divide this into 300, and you end up with 14.2 meters. Round this up to 15 meters to match the closest answer choice. As in 2A-10.7, the calculator keystrokes are: CLEAR 300 ÷ 21.15 = .

2A-10.10 If you are operating your amateur station on 28150 kHz, in what meter band are you operating?
A. 80 meters
B. 40 meters
C. 15 meters
D. 10 meters

ANSWER D: 28,150 kHz is 28.150 MHz. Don't knock yourself out with precise division—just run 28 MHz into 300, and that comes out close to 10 meters. That's the answer. As in 2A-10.7, the calculator keystrokes are: CLEAR 300 ÷ 28.15 = . The 10.7 answer is rounded down to 10.

One question must be from the following 13 questions:

2A-11.1 Who is eligible to obtain a US amateur operator/primary station license?
A. Anyone except a representative of a foreign government
B. Only a citizen of the United States
C. Anyone
D. Anyone except an employee of the United States Government

ANSWER A: There is no nationality requirement to become an amateur operator in the United States. Tests are in English, so the applicant must be able to speak and understand English.

2A-11.2 Who is *not* eligible to obtain a US amateur operator/primary station license?
A. Any citizen of a country other than the United States
B. A representative of a foreign government
C. No one
D. An employee of the United States Government

ANSWER B: Sorry, no government officials from foreign countries are allowed to obtain an amateur operator license.

2A-12.1 What FCC examination elements are required for a Novice Class license?
A. Elements 1(A) and 2(A)
B. Elements 1(A) and 3(A)
C. Elements 1(A) and 2
D. Elements 2 and 4

ANSWER C: Element 1A is the 5-wpm Morse code test. Element 2 is the Novice written examination.

2A-12.2 What is an FCC Element 1(A) examination intended to prove?
A. The applicant's ability to send and receive texts in the international Morse code at not less than 5 words per minute

B. The applicant's ability to send and receive texts in the international Morse code at not less than 13 words per minute
C. The applicant's knowledge of Novice class theory and regulations
D. The applicant's ability to recognize Novice frequency assignments and operating modes

ANSWER A: Sending Morse code is easy on your part. It's a great way to get warmed up before you try the receiving test where you write down the letters. First you send, and then you receive.

2A-12.3 What is an FCC Element 2 examination?
A. A test of the applicant's ability to send and receive Morse code at 5 words per minute
B. The written examination concerning the privileges of a Technician Class operator license
C. A test of the applicant's ability to recognize Novice frequency assignments
D. The written examination concerning the privileges of a Novice Class operator license

ANSWER D: The written examination is a snap. In fact, right now you are memorizing answers that will help you pass the written test.

2A-13.1 Who is eligible to obtain an FCC-issued written authorization for an amateur station?
A. A licensed amateur operator
B. Any unlicensed person, except an agent of a foreign government
C. Any unlicensed person, except an employee of the United States Government
D. Any unlicensed United States Citizen

ANSWER A: You must be a licensed ham operator to obtain a station license.

2A-14.1 Why is an amateur operator required to furnish the FCC with a current mailing address served by the US Postal service?
A. So the FCC has a record of the location of each amateur station
B. In order to comply with the Commission's rules and so the FCC can correspond with the licensee
C. So the FCC can send license-renewal notices
D. So the FCC can compile a list for use in a call sign directory

ANSWER B: When you fill out the FCC Form 610 (one is in the back of this book) to apply for your amateur operator/primary station license, it asks for your mailing address. Where do you want to get the license? At your home? At your office? At a friend's house? It's a hassle to change your mailing address, so make it some spot that you plan to get your mail for the next few years.

2A-15.1 Which one of the following call signs is a valid US amateur call?
A. UA4HAK
B. KBL7766
C. KA9OLS
D. BY7HY

ANSWER C: There are never more than two letters preceding the number in a ham call sign, followed by another letter, or two, or three. Answer B has too many letters, and ends up with numbers—so it's incorrect. And since Answers A and D don't begin with A, K, W, or N, they are wrong, too.

2A-15.2 Which one of the following call signs is a valid US amateur call?
A. CE2FTF
B. G3GVA
C. UA1ZAM
D. AA2Z

ANSWER D: This is obviously the correct answer because it's the only answer that begins with A, K, W, or N—all U.S. prefixes.

2A-15.3 Which one of the following call signs is not a valid US amateur call?
A. KDV5653
B. WA1DVU
C. KA5BUG
D. NT0Z

ANSWER A: Be careful! This is one of those reverse questions where you need to choose the wrong answer. Answer A is your best choice because it contains too many letters and too many numbers to be an Amateur Radio call sign. Looks like an old CB call sign to me. And hey, there's nothing wrong with being a prior CB radio operator. Many CBers make excellent amateur operators.

2A-15.4 What letters may be used for the first letter in a valid US amateur call sign?
A. K, N, U and W
B. A, K, N and W
C. A, B, C and D
D. A, N, V and W

ANSWER B: In the United States, all ham call signs begin with W, A, N or K. This is because, as an aid to enforcement, all transmitting stations are required to identify themselves at intervals when they are in operation. Radio does not respect national boundaries. By international agreement, the first characters of the call sign indicate the country in which the station is authorized to operate. The only prefixes allocated to United States amateur stations are: AA-AL, KA-KZ, NA-NZ, and WA-WZ.

2A-15.5 Excluding special-event call signs that may be issued by the FCC, what numbers may be used in a valid US call sign?
A. Any double-digit number, 10 through 99
B. Any double-digit number, 22 through 45
C. Any single digit, 1 through 9
D. A single digit, 0 through 9

ANSWER D: You can always get a good idea where someone is by the number in their call sign. That is, unless they moved to another state and didn't wish to change their call sign. You can buy colorful charts that illustrate the geographic area assigned to Amateur Radio call sign numbers.

US Call Signs

2A-16.1 Your Novice license was issued on November 1, 1988. When will it expire?
A. On the date specified on the license
B. November 30, 1998
C. November 1, 1993
D. November 1, 1990

ANSWER A: Hams now have a 10-year period for their license before it needs to be renewed. There is a 2-year grace period if you forget to renew—but you can't operate after your license has expired. After the 2-year license expiration grace period, your license is lost, and you need to start all over again. Don't forget to renew!

One question must be from the following 13 questions:

2A-17.1 What does the term *emission* mean?
A. RF signals transmitted from a radio station
B. Signals refracted by the E layer
C. Filter out the carrier of a received signal
D. Baud rate

ANSWER A: Signals that leave your transmitter are called radio frequencies (abbreviated RF). RF signals have a frequency above the audio range limit (about 20 kilohertz). The RF alternating current generates signals that radiate from the antenna through space. The entire range of radio frequency wavelengths is called the *radio spectrum*.

2A-17.2 What emission types are Novice control operators permitted to use on the 80-meter band?
A. CW only
B. Data only
C. RTTY only
D. Phone only

ANSWER A: CW only means Morse code. You may operate telegraphy on 3675-3725 kHz your Novice 80-meter wavelength band privileges. A1A is code in Table 2A-17.2.

Table 2A-17.2 ITU Emission Designators

First symbol – type of main carrier modulation
(Note: Modulation is the process of varying the radio wave to convey information.)
N – Emission of an unmodulated carrier
A – Amplitude Modulation
J – Single sideband, suppressed carrier (the emission you are authorized on ten meters between 28.3 and 28.5 MHz)
F – Frequency modulation
G – Phase modulation
P – Sequence of unmodulated pulses

Second symbol – Nature of signals modulating the main carrier
Ø – No modulating signal
1 – Single channel digital information, no modulation
2 – Single channel digital information with modulation
3 – Single channel carrying analog information
7 – Two or more channels carrying digital information
8 – Two or more channels carrying analog information
9 – Combination of digital and analog information

Third symbol – Type of information to be transmitted
N – No information transmitted
A – Telegraphy to be received by human hearing
B – Telegraphy to be received by automatic equipment
C – Facsimile, transmission of pictures by radio
D – Data transmission, telemetry, telecommand
E – Telephony (voice information)
F – Video, television
W – Combination of above

2A-17.3 What emission types are Novice control operators permitted to use in the 40-meter band?
A. CW only
B. Data only
C. RTTY only
D. Phone only
ANSWER A: Novice hams may use Morse code (CW, A1A) on 40 meters, 7100-7150 kHz.

2A-17.4 What emission types are Novice control operators permitted to use in the 15-meter band?
A. CW only
B. Data only
C. RTTY only
D. Phone only
ANSWER A: Morse code (CW, A1A), from 21,100-21,200 kHz.

2A-17.5 What emission types are Novice control operators permitted to use from 3675 to 3725 kHz?
A. Phone only
B. CW and phone
C. All amateur emission privileges authorized for use on those frequencies
D. CW only
ANSWER D: Novices may use Morse code only on 80 meters.

2A-17.6 What emission types are Novice control operators permitted to use from 7100 to 7150 kHz in ITU Region 2?
A. CW and Data
B. Phone
C. All amateur emission privileges authorized for use on those frequencies
D. CW only

ANSWER D: Your Novice station can transmit continuous wave (CW) or A1A only on 40 meters.

2A-17.7 What emission types are Novice control operators permitted to use on frequencies from 21.1 to 21.2 MHz?
A. CW and data only
B. CW and phone only
C. All amateur emission privileges authorized for use on those frequencies
D. CW only

ANSWER D: On 15 meters, an exceptional long-range band for daytime and evening contacts, Novices are only allowed CW (A1A), Morse Code.

2A-17.8 What emission types are Novice control operators permitted to use on frequencies from 28.1 to 28.3 MHz?
A. All authorized amateur emission privileges
B. Data or phone only
C. CW, RTTY and data
D. CW and phone only

ANSWER C: I wouldn't tell you a fib (F1B) if I said it was okay for Novices to operate digital computers between 28.1 and 28.3. Besides F1B, you can also operate CW (A1A), Morse code. See Table 2A-17.2.

2A-17.9 What emission types are Novice control operators permitted to use on frequencies from 28.3 to 28.5 MHz?
A. All authorized emission privileges
B. CW and data only
C. CW and single-sideband phone only
D. Data and phone only

ANSWER C: Besides Morse code on the voice portion of the Novice band on 10 meters, you may also operate that jiggly voice mode called single sideband, J3E. Just think of the jiggly voice reminding you of the "J" in J3E of Table 2A-17.2.

2A-17.10 What emission types are Novice control operators permitted to use on the amateur 220-MHz band in ITU Region 2?
A. CW and phone only
B. CW and data only
C. Data and phone only
D. All amateur emission privileges authorized for use on 220 MHz

ANSWER D: Within your Novice privileges at 222.1 to 223.91, you may transmit any emission of your choice. That includes television, too! Your repeated *output* may also legally exceed your Novice band limits.

2A-17.11 What emission types are Novice control operators permitted to use on the amateur 1270-MHz band?

A. Data and phone only
B. CW and data only
C. CW and phone only
D. All amateur emission privileges authorized for use on 1270 MHz
ANSWER D: Within your privileges from 1270 to 1295 MHz, you may operate any emission privileges—including television and all forms and speeds of packet.

2A-17.12 On what frequencies in the 10-meter band may a Novice control operator use single-sideband phone?
A. 3700 to 3750 kHz
B. 7100 to 7150 kHz
C. 21100 to 21200 kHz
D. 28300 to 28500 kHz
ANSWER D: There is plenty of long-range excitement on 28.3 to 28.5MHz, 10 meters!

2A-17.13 On what frequencies in the 1.25-meter band in ITU Region 2 may a Novice control operator use FM phone emission?
A. 28.3 to 28.5 MHz
B. 144.0 to 148.0 MHz
C. 222.1 to 223.91 MHz
D. 1240 to 1270 MHz
ANSWER C: Memorize this. These are your transmit privileges on the amateur service 220-MHz band. And good news—as long as you transmit within your own band limits, you may operate through 220-MHz repeaters that are located just outside Novice band limits—legally! So if someone asks you to go to a repeater on a frequency just outside of band limits for Novice, chances are the input is within Novice limits, and that's perfectly legal.

One question must be from the following 13 questions:

2A-18.1 What amount of output transmitting power may a Novice Class control operator use when operating below 30 MHz?
A. 200 watts input
B. 250 watts output
C. 1500 watts PEP output
D. The minimum legal power necessary to carry out the desired communications
ANSWER D: Be careful of this one—the answer is not a numerical one, but rather a philosophical one. Always run the minimum amount of power to make contact with another station.

2A-18.2 What is the maximum transmitting power ever permitted to be used by an amateur station transmitting in the 80-, 40- and 15-meter Novice bands?
A. 75 watts PEP output
B. 100 watts PEP output
C. 200 watts PEP output
D. 1500 watts PEP output
ANSWER C: Here they want to know if you actually understand the maximum power level in watts. Novice bands on 80, 40, and 15 meters are protected from high-power operation by any class of operator. Everyone within your Novice

band limits must reduce power down to 200 watts PEP output or less. This means you won't have to battle a kilowatt station within your CW band privileges.

2A-18.3 What is the maximum transmitting power permitted an amateur station transmitting on 3725 kHz?
A. 75 watts PEP output
B. 100 watts PEP output
C. 200 watts PEP output
D. 1500 watts PEP output
ANSWER C: Same thing here—everyone, including Extra Class operators, must stay below 200 watts PEP output.

2A-18.4 What is the maximum transmitting power permitted an amateur station transmitting on 7125 kHz?
A. 75 watts PEP output
B. 100 watts PEP output
C. 200 watts PEP output
D. 1500 watts PEP output
ANSWER C: Here is another Novice CW sub-band, so every ham in the U.S. may not run more than 200 watts PEP output.

2A-18.5 What is the maximum transmitting power permitted an amateur station transmitting on 21.125 MHz?
A. 75 watts PEP output
B. 100 watts PEP output
C. 200 watts PEP output
D. 1500 watts PEP output
ANSWER C: And yet another Novice sub-band, where under 200 watts PEP output is the legal limit.

2A-19.1 What is the maximum transmitting power permitted an amateur station with a Novice control operator transmitting on 28.125 MHz?
A. 75 watts PEP output
B. 100 watts PEP output
C. 200 watts PEP output
D. 1500 watts PEP output
ANSWER C: As a Novice operator, you are restricted to 200 watts PEP output or less when operating on the 10-meter band.

2A-19.2 What is the maximum transmitting power permitted an amateur station with a Novice control operator transmitting in the amateur 10-meter band?
A. 25 watts PEP output
B. 200 watts PEP output
C. 1000 watts PEP output
D. 1500 watts PEP output
ANSWER B: Although 200 watts output on 10 meters is allowed, try to keep your power output to a minimum. This band could cause second harmonic interference to your neighbors on television Channel 2.

2A-19.3 What is the maximum transmitting power permitted an amateur station with a Novice control operator transmitting in the amateur 220-MHz band?

A. 5 watts PEP output
B. 10 watts PEP output
C. 25 watts PEP output
D. 200 watts PEP output

ANSWER C: On this VHF band, your power output is limited to 25 watts by FCC order. Don't worry about this limitation—almost everybody else on this band runs no more than 25 watts of power anyway! That's plenty of power to work any repeater you can hear.

2A-19.4 What is the maximum transmitting power permitted an amateur station with a Novice control operator transmitting in the amateur 1270-MHz band?
A. 5 milliwatts PEP output
B. 500 milliwatts PEP output
C. 1 watt PEP output
D. 5 watts PEP output

ANSWER D: The 1270-MHz band is in the microwave region. You are approaching frequencies that will cook a turkey in a microwave oven. The FCC limits your power output to 5 watts to keep you from accidentally frying yourself in this region. Most other hams are also operating at the 5 watt level to reduce their exposure to unnecessary amounts of RF radiation.

2A-19.5 What amount of transmitting power may an amateur station with a Novice control operator use in the amateur 220-MHz band?
A. Not less than 5 watts PEP output
B. The minimum legal power necessary to maintain reliable communications
C. Not more than 50 watts PEP output
D. Not more than 200 watts PEP output

ANSWER B: Be careful—they are trying to trick you into answering this question incorrectly with a numerical answer. Always run the minimum legal power necessary for reliable communications. Incidentally, your maximum power output on the 220- MHz band is 25 watts, and none of these answers give you that option.

2A-20.1 What term is used to describe narow-band direct-printing telegraphy emissions?
A. Teleport communications
B. Direct communications
C. RTTY communications
D. Third-party communications

ANSWER C: With the right interface equipment, it will probably take less than 15 minutes to tie in your present ham radio equipment to your present home computer gear. This is RTTY communications using Amateur Radio.

2A-20.2 What term is used to describe telemetry, telecommand and computer communications emissions?
A. Teleport communications
B. Direct communications
C. Data communications
D. Third-party communications

ANSWER C: You don't even need to be in the room to let your computer interact on ham radio. It will faithfully give your call sign in your absence, and rules permit this legal, unattended-type operation.

2A-20.3 On what frequencies in the 10-meter band are Novice control operators permitted to transmit RTTY?
 A. 28.1 to 28.5 MHz
 B. 28.0 to 29.7 MHz
 C. 28.1 to 28.2 MHz
 D. 28.1 to 28.3 MHz
ANSWER D: RTTY stands for radio teletypewriter, and you can transmit and receive radio teletype messages as well as Morse code from 28.1 to 28.3 MHz.

One question must be from the following 10 questions:

2A-21.1 Who is held responsible for the proper operation of an amateur station?
 A. Only the control operator
 B. Only the station licensee
 C. Both the control operator and the station licensee
 D. The person who owns the property where the station is located
ANSWER C: If you are visiting another ham's station, and you operate the equipment with you acting as a control operator, both you and the other ham are jointly responsible for the proper operation of the station.

2A-21.2 You allow another amateur operator to use your amateur station. What are your responsibilities, as the station licensee?
 A. You and the other amateur operator are equally responsible for the proper operation of your station
 B. Only the control operator is responsible for the proper operation of the station.
 C. As the station licensee, you must be at the control point of your station whenever it is operated.
 D. You must notify the FCC when another amateur will be the control operator of your station.
ANSWER A: It's okay to let another ham use your station. However, as the station licensee, you are jointly responsible for the proper operation of the equipment.

2A-21.3 What is your primary responsibility as the station licensee?
 A. You must permit any licensed amateur operator to operate your station at any time upon request.
 B. You must be present whenever the station is operated.
 C. You must notify the FCC in writing whenever another amateur operator will act as the control operator.
 D. You are responsible for the proper operation of the station for which you are licensed.
ANSWER D: When a fellow ham is using your equipment and station call sign, remember your responsibility of making sure everything is legal.

2A-21.4 If you are the licensee of an amateur station when are you *not* responsible for its proper operation?
A. Only when another licensed amateur is the control operator
B. The licensee is responsible for the proper operation of the station for which he or she is licensed.
C. Only after notifying the FCC in writing that another licensed amateur will assume responsibility for the proper operation of your station
D. Only when your station is in repeater operation
ANSWER B: There is never a time when you can get out of being responsible for the transmission from that station.

2A-22.1 When must an amateur station have a control operator?
A. A control operator is only required for training purposes.
B. Whenever the station receiver is operated
C. Whenever the station is transmitting
D. A control operator is not required.
ANSWER C: Every ham station must have a responsible control operator unless it's an automated repeater station under "automatic control." The FCC defines a control operator as: "An amateur operator designated by the licensee of an amateur station to also be responsible for the emissions from that station." *Automatic control* means the use of devices or procedures for control without the control operator being present at the control point when the station is transmitting.

2A-22.2 Another amateur gives you permission to use her amateur station. What are your responsibilities, as the control operator?
A. Both you and she are equally responsible for the proper operation of her station
B. Only the station licensee is responsible for the proper operation of the station, not you the control operator
C. You must be certain the station licensee has given proper FCC notice that you will be the control operator.
D. You must inspect all antennas and related equipment to ensure they are working properly.
ANSWER A: As a control operator, you are both responsible for the transmissions from that station.

2A-23.1 Who may be the control operator of an amateur station?
A. Any person over 21 years of age
B. Any properly licensed amateur operator that is designated by the station licensee
C. Any licensed amateur operator with an Advanced Class license or higher
D. Any person over 21 years of age with a General Class license or higher
ANSWER B: As a Novice operator, you may only assume the control operator responsibility as your Novice license permits. If you are visiting another station operated by a General Class operator on General Class frequencies, your Novice license does not allow you to assume control operator functions.

2A-24.1 Where must an amateur operator be when he or she is performing the duties of control operator?
- A. Anywhere in the same building as the transmitter
- B. At the control point of the amateur station
- C. At the station entrance, to control entry to the room
- D. Within sight of the station monitor, to view the output spectrum of the transmitter

ANSWER B: If you let another ham use your station, your control operator responsibilities require you to stay in the room, right at the radio equipment, supervising the communications.

2A-25.1 Where must you keep your amateur operator license when you are operating a station?
- A. Your original operator license must always be posted in plain view.
- B. Your original operator license must always be taped to the inside front cover of your station log.
- C. You must have the original or a photocopy of your operator license in your possession.
- D. You must have the original or a photocopy of your operator license posted at your primary station location. You need not have the original license nor a copy in your possession to operate another station.

ANSWER C: Whenever you are operating an amateur station, you must keep a copy of your license with you, or posted by the equipment you are operating.

2A-26.1 Where must you keep your written authorization for an amateur station?
- A. Your original station license must always be taped to the inside front cover of your station log.
- B. Your original station license must always be posted in plain view.
- C. You must post the original or a photocopy of your station license at the main entrance to the transmitter building.
- D. The original or a photocopy of the written authorization for an amateur station must be retained at the station

ANSWER D: Keep the original of your license safe and sound at your station. Make plenty of copies for multiple control points. Author's quote: "I have copies of my license in my car, on the boat, and in my airplane."

One question must be from the following 14 questions:

2A-27.1 How often must an amateur station be identified?
- A. At the beginning of the contact and at least every ten minutes during a contact
- B. At least once during each transmission
- C. At least every ten minutes during a contact and at the end of the contact
- D. Every 15 minutes during a contact and at the end of the contact

ANSWER C: Give your call letters regularly, and remember, by law you don't need to give them at the beginning of the transmission. However, it makes good sense to start out with your call letters.

2A-27.2 As an amateur operator, how should you correctly identify your station?
A. With the name and location of the control operator
B. With the station call sign
C. With the call of the control operator, even when he or she is visiting another radio amateur's station
D. With the name and location of the station licensee, followed by the two-letter designation of the nearest FCC Field Office

ANSWER B: Use your own call sign—not someone else's call sign—when operating from your own station.

2A-27.3 What station identification, if any, is required at the beginning of communication?
A. The operator originating the contact must transmit both call signs.
B. No identification is required at the beginning of the contact.
C. Both operators must transmit their own call signs.
D. Both operators must transmit both call signs.

ANSWER B: Isn't that strange! By FCC law, you can actually communicate for nearly 10 minutes without identifying. You must identify at the ten minute point—or at the end of the communication, whichever comes first. Most hams identify when signing on a specific frequency so the other person will know who they are talking to. It is a good idea.

2A-27.4 What station identification, if any, is required at the end of a communication?
A. Both stations must transmit their own call sign, assuming they are FCC-licensed
B. No identification is required at the end of the contact
C. The station originating the contact must always transmit both call signs
D. Both stations must transmit their own call sign followed by a two-letter designator for the nearest FCC field office

ANSWER A: When you sign off, always end with your call sign. The other operator will always end with his or her call sign.

2A-27.5 What do the FCC rules for amateur station identification generally require?
A. Each amateur station shall give its call sign at the beginning of each communication, and every ten minutes or less during a communication.
B. Each amateur station shall give its call sign at the end of each communication, and every ten minutes or less during a communication.
C. Each amateur station shall give its call sign at the beginning of each communication, and every five minutes or less during a communication.
D. Each amateur station shall give its call sign at the end of each communication, and every five minutes or less during a communication.

ANSWER B: You are the only one in the world with your particular call sign. Say it slowly, and with pride. It's yours.

2A-27.6 What is the fewest number of times you must transmit your amateur station identification during a 25 minute QSO?
- A. 1
- B. 2
- C. 3
- D. 4

ANSWER C: Once before minute 10, again before minute 20, and again at sign off. The answer is three times.

2A-27.7 What is the longest period of time during a QSO that an amateur station does not need to transmit its station identification?
- A. 5 minutes
- B. 10 minutes
- C. 15 minutes
- D. 20 minutes

ANSWER B: Some hams use a 10-minute timer to remind them to give their call signs. Part 97 stipulates: "Each amateur shall give its call sign at the end of each communication, and every ten minutes or less during a communication."

2A-28.1 With which amateur stations may an FCC-licensed amateur station communicate?
- A. All amateur stations
- B. All public noncommercial radio stations unless prohibited by the station's government
- C. Only with US amateur stations
- D. All amateur stations, unless prohibited by the amateur's government

ANSWER D: As a licensed amateur operator, you are allowed to communicate with fellow ham radio operators throughout the world. Nearly every country in the world has a ham radio service. The only time you cannot talk to a ham in another country is if it is prohibited by that amateur's government.

2A-28.2 With which non-amateur stations may an FCC-licensed amateur station communicate?
- A. No non-Amateur stations
- B. All such stations
- C. Only those authorized by the FCC
- D. Only those who use the International Morse code

ANSWER C: In some very rare cases, the FCC may allow you to talk to a non-ham station—but usually only in an emergency.

2A-29.1 When must the licensee of an amateur station in portable or mobile operation notify the FCC?
- A. One week in advance if the operation will last for more than 24 hours
- B. FCC notification is not required for portable or mobile operation.
- C. One week in advance if the operation will last for more than a week
- D. One month in advance of any portable or mobile operation

ANSWER B: No longer do you need to notify your local FCC office if you are going to be operating away from the location shown on your license.

2A-29.2 When may you operate your amateur station at a location within the United States, its territories or possessions other than the one listed on your station license?

A. Only during times of emergency

B. Only after giving proper notice to the FCC

C. During an emergency or an FCC-approved emergency preparedness drill

D. Whenever you want to

ANSWER D: Just because your license has a permanent station location on it, don't think for a second that this is the only place you may operate your ham set. You can operate it anywhere in the U.S. and its territories and possessions without notifying the FCC. However, using your portable ham set on a commercial airplane is taboo. Using your ham set on cruise ships requires the permission of the captain.

2A-30.1 When are communications pertaining to business or commercial affairs of any party permitted in the amateur service?

A. Only when the immediate safety of human life or immediate protection of property is threatened

B. There are no rules against conducting business communications in the amateur service.

C. No business communications of any kind are ever permitted in the amateur service.

D. Business communications are permitted between the hours of 9 AM to 5 PM, only on weekdays.

ANSWER A: Your local ham instructor has probably drilled into your memory that business communications are taboo on the ham bands. That's correct. However, during times of emergency, anything goes, including business. If you are providing emergency communications for a fire line, it's perfectly legal to use your handheld to dial up an autopatch and order emergency food rations for the firefighters in action. "Let's see, I would like 2,000 Big Macs..."

2A-30.2 You wish to obtain an application for membership in the American Radio Relay League. When would you be permitted to send an Amateur Radio message requesting the application?

A. At any time, since the ARRL is a not-for-profit organization

B. Never. This would facilitate the commercial affairs of the ARRL

C. Only during normal business hours, between 9 AM and 5 PM

D. At any time, since there are no rules against conducting business communications in the amateur service

ANSWER B: Using a ham set to handle any service from a company, or anything else of a business nature for you or the party you are calling, is not legal.

2A-30.3 On your way home from work you decide to order pizza for dinner. When would you be permitted to use the autopatch on your radio club repeater to order the pizza?

A. At any time, since you will not profit from the communications

B. Only during normal business hours, between 9 AM and 5 PM

C. At any time, since there are no rules against conducting business communications in the amateur service

D. Never. This would facilitate the commercial affairs of a business

ANSWER D: You can't call a restaurant to make a dinner reservation. However, it is okay for you to call your friends on an autopatch and ask them to go to dinner with you.

One question must be from the following 10 questions:

2A-31.1 When may an FCC-licensed amateur operator communicate with an amateur operator in a foreign country?
A. Only when the foreign operator uses English as his primary language
B. All the time, except on 28.600 to 29.700 MHz
C. Only when a third-party agreement exists between the US and the foreign country
D. At any time unless prohibited by either the US or the foreign government

ANSWER D: We may speak with every amateur operator in the world. Most foreign hams speak English as the common ham radio language. We are not prohibited from talking with any foreign ham radio operator at this time.

2A-32.1 When may an amateur station be used to transmit messages for hire?
A. Under no circumstances may an amateur station be hired to transmit messages
B. Modest payment from a non-profit charitable organization is permissible
C. No money may change hands, but a radio amateur may be compensated for services rendered with gifts of equipment or services rendered as a returned favor
D. All payments received in return for transmitting messages by Amateur Radio must be reported to the IRS

ANSWER A: Do not use your ham radio to transmit any messages for hire. Don't confuse ham radio with business radio or cellular telephone services.

2A-32.2 When may the control operator be paid to transmit messages from an amateur station?
A. The control operator may be paid if he or she works for a public service agency such as the Red Cross.
B. The control operator may not be paid under any circumstances.
C. The control operator may be paid if he or she reports all income earned from operating an amateur station to the IRS as receipt of tax-deductible contributions.
D. The control operator may accept compensation if he or she works for a club station during the period in which the station is transmitting telegraphy practice or information bulletins if certain exacting conditions are met.

ANSWER D: Normally you can't get paid to operate a ham station. However, there is one exception. If you are employed by an organization that transmits ham radio news bulletins and CW practice, it is permissible to draw wages for your work.

2A-33.1 When is an amateur operator permitted to broadcast information intended for the general public?
A. Amateur operators are not permitted to broadcast information intended for the general public
B. Only when the operator is being paid to transmit the information

C. Only when such transmissions last less than 1 hour in any 24-hour period
D. Only when such transmissions last longer than 15 minutes

ANSWER A: News bulletins broadcast over the ham radio airwaves must relate solely to Amateur Radio matters or interest the amateur operators tuning in.

2A-34.1 What is third-party communications?

A. A message passed from the control operator of an amateur station to another control operator on behalf of another person
B. Public service communications handled on behalf of a minor political party
C. Only messages that are formally handled through Amateur Radio channels
D. A report of highway conditions transmitted over a local repeater

ANSWER A: Did you know you can let other people talk over your ham set who might not be ham radio operators? That's right, but you must stay right at the microphone to act as a "control operator" and make sure that they abide by the rules. When you link your ham radio into the telephone service, the people you can call would be considered "third-party."

2A-34.2 Who is a third party in amateur communications?

A. The amateur station that breaks into a two-way contact between two other amateur stations
B. Any person for whom a message is passed through amateur communication channels other than the control operators of the two stations handling the message
C. A shortwave listener monitoring a two-way amateur communication
D. The control operator present when an unlicensed person communicates over an amateur station

ANSWER B: When *third-party* traffic is taking place, never leave the room. You must supervise the conversation at all times to insure that the rules concerning illegal messages are being abided by. All messages involving "business interest" and material compensation are prohibited—as well as messages to individuals of certain countries who do not permit Amateur Radio message traffic on behalf of others.

2A-34.3 When is an amateur operator permitted to transmit a message to a foreign country for a third party?

A. Anytime
B. Never
C. Anytime, unless there is a third-party communications agreement between the US and the foreign government
D. When there is a third-party communications agreement between the US and the foreign government, or when the third party is eligible to be the control operator of the station

ANSWER D: The following list is an example of countries with whom we have a third-party agreement. As you will see, most of our third-party agreements are with South American countries, with a few countries to our west, but almost no countries in Europe.

Table 2A-34.3 List of Countries Permitting Third-Party Traffic

V2	Antigua & Barbuda	YS	El Salvador	ZP	Paraguay
LU	Argentina	C5	The Gambia	OA	Peru
VK	Australia	9G	Ghana	VR6	Pitcairn Island*
V3	Belize	J3	Grenada	V4	St Christopher (St. Kitts)
CP	Bolivia	TG	Guatemala		and Nevis
PY	Brazil	8R	Guyana	J6	St. Lucia
VE	Canada	HH	Haiti	J8	St. Vincent
CE	Chile	HR	Honduras	3D6	Swaziland
HK	Colombia	4X	Israel	9Y	Trinidad & Tobago
TI	Costa Rica	6Y	Jamaica	GB	United Kingdom**
CO	Cuba	JY	Jordan	CX	Uruguay
J7	Dominica	EL	Liberia	YV	Venezuela
HI	Dominican Rep.	XE	Mexico	4U1ITU	ITU, Geneva
HC	Ecuador	YN	Nicaragua	4U1VIC	VIC, Geneva
		HP	Panama		

* Informal Temporary
** Limited to special-event stations with callsign prefix GB; GB3 excluded

2A-35.1 Is an amateur station permitted to transmit music?
A. The transmission of music is not permitted in the amateur service.
B. When the music played produces no dissonances or spurious emissions
C. When it is used to jam an illegal transmission
D. Only above 1280 MHz

ANSWER A: Don't sing "Happy Birthday" to a friend over ham radio. Music is not allowed. However, your author is not sure that his singing would really constitute music over the airwaves!

2A-36.1 Is the use of codes or ciphers where the intent is to obscure the meaning permitted during a two-way communication in the amateur service?
A. Codes and ciphers are permitted during ARRL-sponsored contests.
B. Codes and ciphers are permitted during nationally declared emergencies.
C. The transmission of codes and ciphers where the intent is to obscure the meaning is not permitted in the amateur service
D. Codes and ciphers are permitted above 1280 MHz.

ANSWER C: Secret codes are not allowed. It's even considered poor practice to use police-type "ten codes" on the air.

2A-36.2 When is an operator in the amateur service permitted to use abbreviations that are intended to obscure the meaning of the message?
A. Only during ARRL-sponsored contests
B. Only on frequencies above 222.5 MHz
C. Only during a declared communications emergency
D. Abbreviations that are intended to obscure the meaning of the message may never be used in the amateur service

ANSWER D: Common abbreviations known to everyone are legal. An example of this is the Q code "QTH," which means "my location is." (See the Appendix for a list of the common Q codes.)

One question must be from the following 11 questions:

2A-37.1 Under what circumstances, if any, may the control operator cause false or deceptive signals or communications to be transmitted?
A. Under no circumstances
B. When operating a beacon transmitter in a "fox hunt" exercise
C. When playing a harmless "practical joke" without causing interference to other stations that are not involved
D. When you need to obscure the meaning of transmitted information to ensure secrecy

ANSWER A: Going on the air using someone else's call sign is strictly forbidden!

2A-37.2 If an amateur operator transmits the word "MAYDAY" when no actual emergency has occurred, what is this called?
A. A traditional greeting in May
B. An Emergency Action System test transmission
C. False or deceptive signals
D. "MAYDAY" has no significance in an emergency situation

ANSWER C: It's wise not to even utter the word "MAYDAY" in the course of your conversation. Reserve this word for the highest of emergencies over the world-wide bands. On local VHF and UHF repeaters, the equivalent of the worldwide word "MAYDAY" is the phrase "Break, break, break." A triple break signifies a local emergency.

2A-38.1 When may an amateur station transmit unidentified communications?
A. A transmission need not be identified if it is restricted to brief tests not intended for reception by other parties.
B. A transmission need not be identified when conducted on a clear frequency or "dead band" where interference will not occur.
C. An amateur operator may never transmit unidentified communications.
D. A transmission need not be identified unless two-way communications or third-party communications handling are involved.

ANSWER C: Use your call sign often on the air. It's required by law, and you should be proud of your call, too.

2A-38.2 What is the meaning of the term *unidentified radio communications or signals*?
A. Radio communications in which the transmitting station's call sign is transmitted in modes other than CW and voice
B. Radio communications approaching a receiving station from an unknown direction
C. Radio communications in which the operator fails to transmit his or her name and QTH
D. Radio communications in which the station identification is not transmitted

ANSWER D: Jumping into a conversation with a quick comment still requires proper identification with your FCC call letters.

2A-38.3 What is the term used to describe a transmission from an amateur station that does not transmit the required station identification?
A. Unidentified communications or signals
B. Reluctance modulation
C. NØN emission
D. Tactical communication

ANSWER A: Although not required, it's common practice to use your call sign at the beginning of a transmission, too. However, the law does allow you to begin communicating without your call sign for up to 10 minutes. While it's not required at the beginning of a transmission, it's required when you sign off.

2A-39.1 When may an amateur operator willfully or maliciously interfere with a radio communication or signal?
A. You may jam another person's transmissions if that person is not operating in a legal manner.
B. You may interfere with another station's signals if that station begins transmitting on a frequency already occupied by your station.
C. You may never willfully or maliciously interfere with another station's transmissions
D. You may expect, and cause, deliberate interference because it is unavoidable during crowded band conditions.

ANSWER C: Ham radio operators pride themselves in being polite. Deliberate interference is rare, and will not be tolerated. The FCC will respond to jamming complaints from the amateur community. You could lose your amateur operator/ primary station license permanently if found guilty of intentional interference.

2A-39.2 What is the meaning of the term malicious interference?
A. Accidental interference
B. Intentional interference
C. Mild interference
D. Occasional interference

ANSWER B: Ham radio operators self-police the bands, and they all work together to resolve and eliminate interference problems—accidental or malicious.

2A-39.3 What is the term used to describe an Amateur Radio transmission that is intended to disrupt other communications in progress?
A. Interrupted CW
B. Malicious interference
C Transponded signals
D. Unidentified transmissions

ANSWER B: Intentionally transmitting over another station already on the air is not courteous, not legal, and out of the spirit of good ham radio operating.

2A-40.1 As an amateur operator, you receive an *Official Notice of Violation* from the FCC. How promptly must you respond?
A. Within 90 days
B. Within 30 days
C. As specified in the Notice
D. The next day

ANSWER C: Hopefully you'll never receive one. If you do, you must respond within 10 days after receiving the notice of violation.

2A-40.2 If you were to receive a voice distress signal from a station on a frequency outside your operator privileges, what restrictions would apply to assisting the station in distress?
- A. You would not be allowed to assist the station because the frequency of its signals were outside your operator privileges
- B. You would be allowed to assist the station only if your signals were restricted to the nearest frequency band of your privileges
- C. You would be allowed to assist the station on a frequency outside of your operator privileges only if you used international Morse code
- D. You would be allowed to assist the station on a frequency outside of your operator privileges using any means of radiocommunications at your disposal

ANSWER D: In an emergency, anything goes! If you hear someone calling "MAYDAY" on a frequency outside of your normal operating privileges, it's perfectly okay to transmit on any frequency to save someone's life.

2A-40.3 If you were in a situation where normal communication systems were disrupted due to a disaster, what restrictions would apply to essential communications you might provide in connection with the immediate safety of human life?
- A. You would not be allowed to communicate at all except to the FCC Engineer in Charge of the area concerned
- B. You would be restricted to communications using only the emissions and frequencies authorized to your operator privileges
- C. You would be allowed to communicate on frequencies outside your operator privileges only if you used international Morse code
- D. You would be allowed to use any means of radiocommunication at your disposal

ANSWER D: In an emergency, use your radio on any frequency, and on any band, and on any service, to reach out for help.

Subelement 2B – Operating Procedures (2 examination questions from 48 questions in 2B)

One question must be from the following 36 questions:

2B-1-1.1 What is the most important factor to consider when selecting a transmitting frequency within your authorized subband?
- A. The frequency should not be in use by other amateurs.
- B. You should be able to hear other stations on the frequency to ensure that someone will be able to hear you.
- C. Your antenna should be resonant at the selected frequency.
- D. You should ensure that the SWR on the antenna feed line is high enough at the selected frequency.

ANSWER A: Always listen for a few seconds before initiating a transmitted call, on your privileges, within the ARRL suggested band plan, and clear of an ongoing conversation.

2B-1-1.2 You wish to contact an amateur station more than 1500 miles away on a summer afternoon. Which band is most likely to provide a successful contact?

A. The 80- or 40-meter wavelength bands
B. The 40- or 15-meter wavelength bands
C. The 15- or 10-meter wavelength bands
D. The 1¼-meter or 23-centimeter wavelength bands

ANSWER C: 10 and 15 meters are the best long-range bands. Use them in the direction of the sun because short wavelength bands follow the sun.

2B-1-1.3 How can on-the-air transmitter tune-up be kept as short as possible?

A. By using a random wire antenna
B. By tuning up on 40 meters first, then switching to the desired band
C. By tuning the transmitter into a dummy load
D. By using twin lead instead of coaxial-cable feed lines

ANSWER C: It is sometimes necessary to listen to your own signal with a companion receiver to make adjustments to the transmitter. Instead of putting the signal out on the air, use a dummy load to absorb your signal and to keep it from causing interference during your testing procedures. This may be a simple light bulb, or a 50-ohm non-inductive resistor capable of handling your power output.

2B-1-2.1 You are having a QSO with your uncle in Pittsburgh when you hear an emergency call for help on the frequency you are using. What should you do?

A. Inform the station that the frequency is in use.
B. Direct the station to the nearest emergency net frequency.
C. Call your local Civil Preparedness Office and inform them of the emergency.
D. Immediately stand by to copy the emergency communication.

ANSWER D: An emergency call always has the highest priority. Do what you can to take down the message accurately, and then call the proper authorities.

2B-2-1.1 What is the format of a standard Morse code CQ call?

A. Transmit the procedural signal "CQ" three times, followed by the procedural signal "DE", followed by your call three times.
B. Transmit the procedural signal "CQ" three times, followed by the procedural signal "DE", followed by your call one time.
C. Transmit the procedural signal "CQ" ten times, followed by the procedural signal "DE", followed by your call one time.
D. Transmit the procedural signal "CQ" continuously until someone answers your call.

ANSWER A: Practice sending "CQ" on a Morse code oscillator before actually going on the air. Don't send it too fast. A response will be at the same rate that you send "CQ."

2B-2-1.2 How should you answer a Morse code CQ call?

A. Send your call sign four times
B. Send the other station's call sign twice, followed by the procedural signal "DE", followed by your call sign twice
C. Send the other station's call sign once, followed by the procedural signal "DE", followed by your call sign four times
D. Send your call sign followed by your name, station location and a signal report

ANSWER B: When you respond to a Morse code "CQ," respond at the rate that the sending station was using. If they were sending fast, then it's okay to send code back to them fast. The letters "QRS" mean "please send more slowly."

2B-2-2.1 At what telegraphy speed should a "CQ" message be transmitted?

A. Only speeds below five WPM
B. The highest speed your keyer will operate
C. Any speed at which you can reliably receive
D. The highest speed at which you can control the keyer

ANSWER C: We use CQ for a general call to communicate with anyone, about anything, on worldwide frequencies. When using telegraphy, a fast CW CQ call will result in a very fast reply since it will be assumed that you can receive Morse code as fast as you can send it. Don't transmit CW faster than you can receive.

2B-2-3.1 What is the meaning of the Morse code character AR?

A. Only the called station transmit
B. All received correctly
C. End of transmission
D. Best regards

ANSWER C: When sending code (A1A), it's much easier to send abbreviations. Be sure to be familiar with these code abbreviations on both the written exam as well as your code test. When you first start on the air, "QRS" can be very important. Other common telegraphy abbreviations are QRX (wait), QST (bulletin to follow), WX (weather) and CUL (see you later). (See the Appendix for a list of common abbreviations).

2B-2-3.2 What is the meaning of the Morse code character SK?

A. Received some correctly
B. Best regards
C. Wait
D. End of contact

ANSWER D: Telegraphy means using code with a telegraph key. Don't confuse this word with telephony which means using a microphone, like a telephone. In telegraphy, the "K" means the same as "over" or "go ahead." The "SK" indicates the sender is signing off or "goodbye."

2B-2-3.3 What is the meaning of the Morse code character BT?

A. Double dash "="
B. Fraction bar "/"
C. End of contact
D. Back to you

ANSWER A: The double dash is a great way to *pause* as you collect your thoughts on CW.

2B-2-3.4 What is the meaning of the Morse code character DN?

A. Double dash "="
B. Fraction bar "/"
C. Done now (end of contact)
D. Called station only transmit

ANSWER B: Another name for "DN" is "slant bar," and some hams may use it when describing the weather. . . "hot/muggy."

2B-2-3.5 What is the meaning of the Morse code character KN?
  A. Fraction bar "/"
  B. End of contact
  C. Called station only transmit
  D. Key now (go ahead to transmit)
ANSWER C: The letters "KN," run together, mean that you are only expecting a response from a directed station on the air—not everyone. Some hams call it "back to you only."

2B-2-4.1 What is the procedural signal "CQ" used for?
  A. To notify another station that you will call on the quarter hour
  B. To indicate that you are testing a new antenna and are not listening for another station to answer
  C. To indicate that only the called station should transmit
  D. A general call when you are trying to make a contact
ANSWER D: CQ is a fun way to establish communications on worldwide bands. We don't use CQ on the VHF or UHF bands. Things are not that formal up there, so use CQ only on the 10-meter, 15-meter, 40-meter, and 80-meter Novice bands.

2B-2-4.2 What is the procedural signal "DE" used for?
  A. To mean "from" or "this is," as in "W9NGT de N9BTT"
  B. To indicate directional emissions from your antenna
  C. To indicate "received all correctly"
  D. To mean "calling any station"
ANSWER A: Abbreviations are much easier to send when working CW (A1A). Abbreviations are very important for you to know for your examinations, both written and code.

2B-2-4.3 What is the procedural signal "K" used for?
  A. To mean "any station transmit"
  B. To mean "all received correctly"
  C. To mean "end of message"
  D. To mean "called station only transmit"
ANSWER A: *Telegraphy* means using code with a telegraph key. Don't confuse this word with *telephony* which means using a microphone, like a telephone. In telegraphy, the "K" means the same as "over" or "go ahead." The "SK" indicates the sender is signing off or "goodbye."

2B-2-5.1 What does the R in the RST signal report mean?
  A. The recovery of the signal
  B. The resonance of the CW tone
  C. The rate of signal flutter
  D. The readability of the signal
ANSWER D: It's up to you and your ear to determine readability on a scale of 1 to 5. Five is best, 3 is fair, and 1 is for the signal that you can just barely make out.

2B-2-5.2 What does the S in the RST signal report mean?
  A. The scintillation of a signal
  B. The strength of the signal
  C. The signal quality
  D. The speed of the CW transmission

ANSWER B: *Signal strength.* The *RST system* is a signal reporting system based on three numerical designators—Readability, Signal Strength, and Tone—as follows:

Readability
1 - Unreadable
2 - Barely readable; occasional words distinguishable.
3 - Readable with considerable difficulty
4 - Readable with practically no difficulty
5 - Perfectly readable.

Signal Strength
1 - Faint and barely perceptible signals
2 - Very weak signals
3 - Weak signals
4 - Fair signals
5 - Fairly good signals
6 - Good signals
7 - Moderately strong signals
8 - Strong signals
9 - Extremely strong signals

Tone
1 - Very rough, broad signals, 60 cycle AC may be present
2 - Very rough AC tone, harsh, broad
3 - Rough, low pitched AC tone, no filtering
4 - Rather rough AC tone, some trace of filtering
5 - Filtered rectified AC note, musical, ripple modulated
6 - Slight trace of filtered tone but with ripple modulation
7 - Near DC tone but trace of ripple modulation
8 - Good DC tone, may have slight trace of modulation
9 - Purest, perfect DC tone with no trace of ripple or modulation.

Source: ARRL

If the signal has the stability characteristic of being crystal controlled, it is sometimes customary, but not in widespread use, to add the suffix letter "X" to the RST telegraphy report. A suffix letter of "C" indicates "chirp," and "K" indicates that a "key click" is present.

2B-2-5.3 What does the T in the RST signal report mean?
A. The tone of the signal
B. The closeness of the signal to "telephone" quality
C. The timing of the signal dot to dash ratio
D. The tempo of the signal

ANSWER A: Only for CW do we use the "T" for tone. If the tone is pure, give it a 9. If you detect a hum on the signal, give it a 5. If it's the worst signal you've ever heard, give it a 1.

2B-2-6.1 What is one meaning of the Q signal "QRS"?
A. Interference from static
B. Send more slowly
C. Send RST report
D. Radio station location is

ANSWER B: When you first start on the air, "QRS" can be very important because it means "send more slowly." Remember, in your CW (A1A) transmissions, it's much easier to send abbreviations; so learn them well, especially from your examinations.

2B-2-6.2 What is one meaning of the Q signal "QRT"?
A. The correct time is
B. Send RST report
C. Stop sending
D. Send more slowly

ANSWER C: Example: It may be time for me to "QRT" for dinner.

2B-2-6.3 What is one meaning of the Q signal "QTH"?
    A. Time here is
    B. My name is
    C. Stop sending
    D. My location is ...
ANSWER D: Example: What is your QTH?

2B-2-6.4 What is one meaning of the Q signal "QRZ," when it is followed with a question mark?
    A. Who is calling me?
    B. What is your radio zone?
    C. What time zone are you in?
    D. Is this frequency in use?
ANSWER A: Call "CQ" on worldwide frequencies to make a contact. If you hear someone calling you and you can't get their call sign, ask for a repeat of their call sign by simply sending in code: QRZ?. In voice, simply say your call sign followed by "QRZ."

2B-2-6.5 What is one meaning of the Q signal "QSL," when it is followed with a question mark?
    A. Shall I send you my log?
    B. Can you acknowledge receipt (of my message)?
    C. Shall I send more slowly?
    D. Who is calling me?
ANSWER B: Hams exchange colorful postcard QSL's to confirm contacts. Always take a look at another ham's QSL card collection. You will find it fascinating!

2B-3-1.1 What is the format of a standard radiotelephone CQ call?
    A. Transmit the phrase "CQ" at least ten times, followed by "this is," followed by your call sign at least two times.
    B. Transmit the phrase "CQ" at least five times, followed by "this is," followed by your call sign once.
    C. Transmit the phrase "CQ" three times, followed by "this is," followed by your call sign three times.
    D. Transmit the phrase "CQ" at least ten times, followed by "this is," followed by your call sign once.

ANSWER C: We use "CQ" as a general call only on the worldwide bands. On the VHF and UHF bands, we are less formal, and simply announce ourselves as being on the air by just giving our call sign.

2B-3-1.2 How should you answer a radiotelephone CQ call?

A. Transmit the other station's call sign at least ten times, followed by "this is," followed by your call sign at least twice.

B. Transmit the other station's call sign at least five times phonetically, followed by "this is," followed by your call sign at least once.

C. Transmit the other station's call sign at least three times, followed by "this is," followed by your call sign at least five times phonetically.

D. Transmit the other station's call sign once, followed by "this is," followed by your call sign given phonetically.

ANSWER D: If you hear a lively CQ call, then go ahead and respond to that station and enjoy a great conversation. If the station is coming in clear, you would only need to call it once, followed by "This is," and then give your call sign, slowly, and phonetically. Your author also likes to give his location.

2B-3-2.1 How is the call sign "KA3BGQ" stated in Standard International Phonetics?

A. Kilo Alfa Three Bravo Golf Quebec

B. King America Three Bravo Golf Quebec

C. Kilowatt Alfa Three Bravo George Queen

D. Kilo America Three Baker Golf Quebec

ANSWER A: When you voice your call sign, a "P" might sound like a "B", and that sounds like a "C", or maybe it was "D". To end the confusion, use the phonetic alphabet the first time you make contact with a new station. The standard *International Telecommunication Union Phonetic Alphabet* is:

A	-	Alpha	H	-	Hotel	O - Oscar	V - Victor
B	-	Bravo	I	-	India	P - Papa	W - Whiskey
C	-	Charlie	J	-	Juliette	Q - Quebec	X - X-Ray
D	-	Delta	K	-	Kilo	R - Romeo	Y - Yankee
E	-	Echo	L	-	Lima	S - Sierra	Z - Zulu
F	-	Foxtrot	M	-	Mike	T - Tango	
G	-	Golf	N	-	November	U - Uniform	

2B-3-2.2 How is the call sign "WE5TZD" stated phonetically?

A. Whiskey Echo Foxtrot Tango Zulu Delta

B. Washington England Five Tokyo Zanzibar Denmark

C. Whiskey Echo Five Tango Zulu Delta

D. Whiskey Easy Five Tear Zebra Dog

ANSWER C: It's almost guaranteed you will get at least one question relating to the phonetic alphabet on your upcoming examination. Memorize these phonetics, and you will have no problem figuring out the right answer.

2B-3-2.3 How is the call sign "KC4HRM" stated phonetically?

A. Kilo Charlie Four Hotel Romeo Mike

B. Kilowatt Charlie Four Hotel Roger Mexico

C. Kentucky Canada Four Honolulu Radio Mexico

D. King Charlie Foxtrot Hotel Roger Mary

ANSWER A: Memorize those phonetics.

2B-3-2.4 How is the call sign "AF6PSQ" stated phonetically?

A. America Florida Six Portugal Spain Quebec
B. Adam Frank Six Peter Sugar Queen
C. Alfa Fox Sierra Papa Santiago Queen
D. Alfa Foxtrot Six Papa Sierra Quebec

ANSWER D: If you don't memorize your phonetics, you're going to blow valuable points on your test.

2B-3-2.5 How is the call sign "NB8LXG" stated phonetically?

A. November Bravo Eight Lima Xray Golf
B. Nancy Baker Eight Love Xray George
C. Norway Boston Eight London Xray Germany
D. November Bravo Eight London Xray Germany

ANSWER A: Can you imagine flunking the test because you didn't spend a few minutes memorizing your phonetics?

2B-3-2.6 How is the call sign "KJ1UOI" stated phonetically?

A. King John One Uncle Oboe Ida
B. Kilowatt George India Uncle Oscar India
C. Kilo Juliette One Uniform Oscar India
D. Kentucky Juliette One United Ontario Indiana

ANSWER C: Every good ham knows the phonetic alphabet—memorize it.

2B-3-2.7 How is the call sign "WV2BPZ" stated phonetically?

A. Whiskey Victor Two Bravo Papa Zulu
B. Willie Victor Two Baker Papa Zebra
C. Whiskey Victor Tango Bravo Papa Zulu
D. Willie Virginia Two Boston Peter Zanzibar

ANSWER A: Practice your phonetics. Try writing them down to memorize them.

2B-3-2.8 How is the call sign "NY3CTJ" stated phonetically?

A. Norway Yokohama Three California Tokyo Japan
B. Nancy Yankee Three Cat Texas Jackrabbit
C. Norway Yesterday Three Charlie Texas Juliette
D. November Yankee Three Charlie Tango Juliette

ANSWER D: Practice your phonetics. Say them backwards and forwards.

2B-3-2.9 How is the call sign "KG7DRV" stated phonetically?

A. Kilo Golf Seven Denver Radio Venezuela
B. Kilo Golf Seven Delta Romeo Victor
C. King John Seven Dog Radio Victor
D. Kilowatt George Seven Delta Romeo Video

ANSWER B: Will this one be on your exam? Maybe. Practice your phonetics.

2B-3-2.10 How is the call sign "WX9HKS" stated phonetically?

A. Whiskey Xray Nine Hotel Kilo Sierra
B. Willie Xray November Hotel King Sierra
C. Washington Xray Nine Honolulu Kentucky Santiago
D. Whiskey Xray Nine Henry King Sugar

ANSWER A: Look at those incorrect answers—pretty close to the real things, aren't they? Practice your phonetics.

2B-3-2.11 How is the call sign "AE0/LQY" stated phonetically?

A. Able Easy Zero Lima Quebec Yankee
B. Arizona Ecuador Zero London Queen Yesterday
C. Alfa Echo Zero Lima Quebec Yankee
D. Able Easy Zero Love Queen Yoke

ANSWER C: Be careful that you don't read just the first word of the answer; some start out right, but end wrong. Remember, practice makes perfect.

One question must be from the following 12 questions:

2B-4-1.1 What is the format of a standard RTTY CQ call?

A. Transmit the phrase "CQ" three times, followed by "DE", followed by your call sign two times.
B. Transmit the phrase "CQ" three to six times, followed by "DE," followed by your call sign three times.
C. Transmit the phrase "CQ" ten times, followed by the procedural signal "DE", followed by your call one time.
D. Transmit the phrase "CQ" continuously until someone answers your call.

ANSWER B: Radioteletype, or RTTY, operation involves typewriters actuated by binary-coded radio signals. More home computers are used today by hams to receive RTTY than teleprinters. We send CQ six times, rather than three times.

2B-4-2.1 You receive an RTTY CQ call at 45 bauds. At what speed should you respond?

A. 22-1/2 bauds
B. 45 bauds
C. 90 bauds
D. Any speed, since radioteletype systems adjust to any signal rate

ANSWER B: Operating an RTTY station with your home computer is fun. It's also challenging. Not only do you need to tune in the station correctly, but you also need to match the speed of your equipment with the sending rate of the station on the air. 45 bauds is common on worldwide bands.

Station Terminal

2B-5-1.1 What does the term *connected* mean in a packet-radio link?

A. A telephone link has been established between two amateurs.
B. An Amateur Radio message has reached the station for local delivery.
C. The transmitting station is sending data specifically addressed to the receiving station, and the receiving station is acknowledging that the data has been received correctly.
D. The transmitting station and a receiving station are using a certain digipeater, so no other contacts can take place until they are finished.

ANSWER C: Your packet radio system gives you many winking lights that indicate your station is interacting with another station.

2B-5-1.2 What does the term *monitoring* mean on a frequency used for packet radio?

A. The FCC is copying all messages to determine their content.
B. A member of the Amateur Auxiliary to the FCC's Field Operations Bureau is copying all messages to determine their content.
C. The receiving station's video monitor is displaying all messages intended for that station, and is acknowledging correct receipt of the data.
D. The receiving station is displaying information that may not be addressed to that station, and is not acknowledging correct receipt of the data.

ANSWER D: Eavesdropping is a favorite practice on packet radio—and no one will know that you are there.

2B-5-2.1 What is a digipeater?

A. A packet-radio station used to retransmit data that is specifically addressed to be retransmitted by that station
B. An Amateur Radio repeater designed to retransmit all audio signals in a digital form
C. An Amateur Radio repeater designed using only digital electronics components
D. A packet-radio station that retransmits any signals it receives

ANSWER A: Long-time ham radio operators have established excellent repeater stations and packet repeater set-ups for everyone to use. We strongly support those hams that give of their time to help you extend your communications range, and we hope that you will support their stations, too, by joining their membership organizations.

2B-5-2.2 What is the meaning of the term network in packet radio?

A. A system of telephone lines interconnecting packet-radio stations to transfer data
B. A method of interconnecting packet-radio stations so that data can be transferred over long distances
C. The interlaced wiring on a terminal-node-controller board
D. The terminal-node-controller function that automatically rejects another caller when the station is connected

ANSWER B: Packet stations are hooking up throughout the country into networks covering many states. It may take you a few weeks to figure out the system, but your new enhanced Novice license gives you full privileges to take advantage of this new communications medium.

2B-6-1.1 What is a good way to establish a contact on a repeater?
A. Give the call sign of the station you want to contact three times
B. Call the other operator by name and then give your call sign three times
C. Call the desired station and then identify your own station
D. Say, "Breaker, breaker," and then give your call sign

ANSWER C: Before taking advantage of your new repeater privileges, do a lot of listening. This will help you develop good operating technique. Don't use any jargon that you may have used in other radio services. It may take a few weeks to shake your old lingo, but hams have their own new lingo that you will want to learn and use. Simply stating your call sign is a good opener in any situation and on any frequency.

2B-6-2.1 What is the main purpose of a repeater?
A. To provide a station that makes local information available 24 hours a day
B. To provide a means of linking amateur stations with the telephone system
C. To retransmit NOAA weather information during severe storm warnings
D. Repeaters extend the operating range of portable and mobile stations

ANSWER D: A repeater (or "machine" as hams commonly call it) retransmits low-powered handheld, base station or mobile radio signals from tall high-gain antennas at higher power thereby extending communications range. Repeater communications is the ham's very popular social party line! You have repeater privileges on the 220- and 1270-MHz bands. You should purchase a repeater guide so that you can learn the frequencies of the various repeaters in your area. While Novice may operate through 220-MHz and 1270-MHz repeaters, they may not establish or be the control operator of a repeater.

2B-6-3.1 What does it mean to say that a repeater has an input and an output frequency?
A. The repeater receives on one frequency and transmits on another.
B. All repeaters offer a choice of operating frequency, in case one is busy.
C. One frequency is used to control repeater functions and the other frequency is the one used to retransmit received signals.
D. Repeaters require an access code to be transmitted on one frequency while your voice is transmitted on the other.

ANSWER A: Repeater directories publish the repeater frequencies by output. The plus (+) or minus (−) indicates the input "split" that you dial in on your VHF or UHF ham set. A plus (+) indicates a higher input and a minus (−) indicates a lower input. When you start to transmit, your transmitter should automatically go to the proper input frequency. Some repeaters also require a subaudible tone as part of your input transmission. Ask the local operators how to engage the tone signal on your ham radio set.

2B-6-4.1 When should simplex operation be used instead of using a repeater?
A. Whenever greater communications reliability is needed
B. Whenever a contact is possible without using a repeater
C. Whenever you need someone to make an emergency telephone call
D. Whenever you are traveling and need some local information

ANSWER B: If you are communicating with a station that is located within 10 miles of you, go "simplex" (or direct) rather than through a repeater. This localizes your transmissions and frees the repeater for more distant contacts.

2B-6-5.1 What is an autopatch?

A. A repeater feature that automatically selects the strongest signal to be repeated
B. An automatic system of connecting a mobile station to the next repeater as it moves out of range of the first
C. A device that allows repeater users to make telephone calls from their portable or mobile stations
D. A system that automatically locks other stations out of the repeater when there is a QSO in progress

ANSWER C: Now you have a car phone—and a pocket phone! Many ham repeaters are tied into automatic telephone system interconnections. Once you join a repeater group, you may be given the specific access code for making local phone calls. Remember, it's not legal to call the office or make business phone calls from an Amateur Radio set.

Don't confuse your new pleasure-telephone patch capabilities with the utility of a regular cellular telephone. For business, you should use cellular telephone or CB. For strictly personal phone calls, use your ham autopatch, and be aware that the whole world is listening! Phone patches are anything but private.

2B-6-5.2 What is the purpose of a repeater time-out timer?

A. It allows the repeater to have a rest period after heavy use.
B. It logs repeater transmit time to determine when the repeater mean time between failure rating is exceeded.
C. It limits repeater transmission time to no more than ten minutes.
D. It limits repeater transmission time to no more than three minutes.

ANSWER D: Time-out timers keep repeater operators from getting long-winded. The repeater cycles off the air if it doesn't get a few seconds break during one long transmission. Some timers are as short as 30 seconds. It's always good practice to keep your transmissions shorter than a one-half minute period. If you need to talk longer, announce, "Reset," release the mike button, and let the repeater reestablish its time-out timer.

Subelement 2C – Radio-Wave Propagation (1 examination question from 18 questions in 2C)

One question must be from the following 18 questions:

2C-1.1 What type of radio-wave propagation occurs when the signal travels in a straight line from the transmitting antenna to the receiving antenna?

A. Line-of-sight propagation
B. Straight-line propagation
C. Knife-edge diffraction
D. Tunnel propagation

ANSWER A: Signals on your 220 and 1270 MHz bands travel in straight lines. Atmospheric refraction may cause the signals to extend 20 percent beyond the optical horizon.

2C-1.2 What path do radio waves usually follow from a transmitting antenna to a receiving antenna at VHF and higher frequencies?
A. A bent path through the ionosphere
B. A straight line
C. A great circle path over either the north or south pole
D. A circular path going either east or west from the transmitter
ANSWER B: These line-of-sight signals bounce off of buildings, and many times will even transmit into elevators. While signals travel in straight lines, they often take some great bounces. They also are bent by local still-air atmospheric conditions.

2C-2.1 What type of propagation involves radio signals that travel along the surface of the Earth?
A. Sky-wave propagation
B. Knife-edge diffraction
C. E-layer propagation
D. Ground-wave propagation
ANSWER D: All stations on all frequencies emit ground waves that hug the earth. They travel out from your transmitter antenna up to approximately 100 miles.

2C-2.2 What is the meaning of the term *ground-wave propagation*?
A. Signals that travel along seismic fault lines
B. Signals that travel along the surface of the earth
C. Signals that are radiated from a ground-plane antenna
D. Signals that are radiated from a ground station to a satellite
ANSWER B: Ground waves are not affected by solar conditions, so you can always depend on ground waves for predictable communications to a nearby station.

2C-2.3 Two amateur stations a few miles apart and separated by a low hill blocking their line-of-sight path are communicating on 3.725 MHz. What type of propagation is probably being used?
A. Tropospheric ducting
B. Ground wave
C. Meteor scatter
D. Sporadic E
ANSWER B: All signals leaving an antenna hug the horizon. These are called ground waves, and usually extend out from the transmitter antenna from 25 miles to 100 miles.

2C-2.4 When compared to sky-wave propagation, what is the usual effective range of ground-wave propagation?
A. Much smaller
B. Much greater
C. The same
D. Dependent on the weather
ANSWER A: Ground waves have a much smaller range than skywaves reflected off the ionosphere. It's quite common for skywave signals on 10 meters located 3,000 miles away to overpower local signals only a few miles away! You will be fascinated with 10-meter propagation.

2C-3.1 What type of propagation uses radio signals refracted back to earth by the ionosphere?
  A. Sky wave
  B. Earth-moon-earth
  C. Ground wave
  D. Tropospheric
ANSWER A: What you couldn't do on CB is a way of life on ham radio! Radio waves are bounced off of the ionospheric E or F layer and may return to earth thousands of miles away! This 10-meter propagation occurs in daylight and evening hours—seldom beyond 9:00 p.m. The ionosphere becomes mirror-like due to ultraviolet radiation given off by the sun. The mirror may be so intense during daylight hours that 10-meter signals may travel half-way around the world. DXing will be fun for you as a Novice!

2C-3.2 What is the meaning of the term *sky-wave propagation*?
  A. Signals reflected from the moon
  B. Signals refracted by the ionosphere
  C. Signals refracted by water-dense cloud formations
  D. Signals retransmitted by a repeater
ANSWER B: On the 10-meter band, skywaves will give you some phenomenal long-range contacts during daylight and twilight hours. Sorry folks, no skywaves on the 220- or 1270-MHz band.

2C-3.3 What does the term *skip* mean?
  A. Signals are reflected from the moon.
  B. Signals are refracted by water-dense cloud formations.
  C. Signals are retransmitted by repeaters.
  D. Signals are refracted by the ionosphere.
ANSWER D: The 10-meter ham band is the only band that has the potential to allow worldwide Novice voice skip communication. Frequencies in the 220-MHz and 1270-MHz region have wavelengths too short for ionospheric refraction—so no skip will ever be found on these bands.

2C-3.4 What is the area of weak signals between the ranges of ground waves and the first hop called?
  A. The skip zone
  B. The hysteresis zone
  C. The monitor zone
  D. The transequatorial zone
ANSWER A: Sometimes skywaves may skip completely over stations a few hundred miles away. This means you won't be able to talk to a station 400 miles away, but will be able to communicate freely with a station 2,000 miles away! This area of no-reception is found within the skip zone. Welcome to ham radio!

2C-3.5 What is the meaning of the term *skip zone?*?
  A. An area covered by skip propagation
  B. The area where a satellite comes close to the earth, and skips off the ionosphere
  C. An area that is too far for ground-wave propagation, but too close for skip propagation
  D. The area in the atmosphere that causes skip propagation

ANSWER C: Increasing your power won't help establish communications with a station that is being missed within your skip zone. If you switch to other bands, chances are lower frequencies (longer wavelength) may skip in closer.

2C-3.6 What type of radio wave propagation makes it possible for amateur stations to communicate long distances?
   A. Direct-inductive propagation
   B. Knife-edge diffraction
   C. Ground-wave propagation
   D. Sky-wave propagation

ANSWER D: The 10-meter band will be full of skip activity as the sunspot activity increases to a peak around 1989 to 1991—the predicted peak of solar cycle 22. Expect short skip signals daily up to 1,000 miles away during the summertime. During the rest of the year, expect almost daily band openings from 2,000 miles to throughout the world during daylight and evening hours. Dawn and dusk are the best times for contacting stations in foreign lands on 10 meters. Have fun!

2C-4.1 How long is an average sunspot cycle?
   A. 2 years
   B. 5 years
   C. 11 years
   D. 17 years

ANSWER C: The sun exhibits periods of high solar activity, peaking every 11 years. During the peak of the sunspot cycles, your 10-meter voice privileges will reach out to worldwide capabilities. At the minimum of the next sunspot cycle (around 1995), solar activity will drop off, and 10-meter long range contacts may only be a summertime phenomenon.

2C-4.2 What is the term used to describe the long-term variation in the number of visible sunspots?
   A. The 11-year cycle
   B. The solar magnetic flux cycle
   C. The hysteresis count
   D. The sunspot cycle

ANSWER D: Never look directly at the sun to detect sunspots. Let a friend that knows astronomy show you how to safely display the image of the sun on a piece of cardboard. You can easily see the sunspots, safely, using safe observation techniques.

2C-5.1 What effect does the number of sunspots have on the maximum usable frequency (MUF)?
   A. The more sunspots there are, the higher the MUF will be.
   B. The more sunspots there are, the lower the MUF will be.
   C. The MUF is equal to the square of the number of sunspots.
   D. The number of sunspots affects the lowest usable frequency (LUF) but not the MUF.

ANSWER A: During periods of low sunspot activity, the maximum usable frequency hovers around 20 MHz. This is below the 10-meter band, so you won't get skip. During periods of multiple sunspots, the MUF will be well above 30 MHz, giving you great skip on your 28-MHz, 10-meter, Novice voice ham band.

2C-5.2 What effect does the number of sunspots have on the ionization level in the atmosphere?
  A. The more sunspots there are, the lower the ionization level will be.
  B. The more sunspots there are, the higher the ionization level will be.
  C. The ionization level of the ionosphere is equal to the square root of the number of sunspots.
  D. The ionization level of the ionosphere is equal to the square of the number of sunspots.
ANSWER B: Listen to WWV, 10 or 15 MHz, at 18 minutes past every hour. They will report on the amount of solar activity, and you can many times predict whether 10 meters is going to be hot, or not, the next day.

2C-6.1 Why can a VHF or UHF radio signal that is transmitted toward a mountain often be received at some distant point in a different direction?
  A. You can never tell what direction a radio wave is traveling in.
  B. These radio signals are easily bent by the ionosphere.
  C. These radio signals are easily reflected by objects in their path.
  D. These radio signals are sometimes scattered in the ectosphere.
ANSWER C: While 220-MHz and 1270-MHz signals never bounce off of the ionosphere, they can be reflected off of local mountains, inversion layers, tall buildings—even off the moon! You can bounce your signal just about anywhere you want with a directional antenna.

2C-6.2 Why can the direction that a VHF or UHF radio signal is traveling be changed if there is a tall building in the way?
  A. You can never tell what direction a radio wave is traveling in.
  B. These radio signals are easily bent by the ionosphere.
  C. These radio signals are easily reflected by objects in their path.
  D. These radio signals are sometimes scattered in the ectosphere.
ANSWER C: Your small 220-MHz or 1270-MHz handheld transceiver will work nicely inside buildings and even in elevators. Signals are so reflective at these frequencies that they manage to find a way out to that distant repeater. You may also find that there are different areas in your house or office that are better than others for working distant repeaters because of nearby reflections.

Subelement 2D – Amateur Radio Practice (4 examination questions from 45 questions in 2D)

One question must be from the following 10 questions:

2D-1.1 How can you prevent the use of your amateur station by unauthorized persons?
  A. Install a carrier-operated relay in the main power line
  B. Install a key-operated "ON/OFF" switch in the main power line
  C. Post a "Danger - High Voltage" sign in the station
  D. Install ac line fuses in the main power line
ANSWER B: You are responsible for unauthorized use of your station even though the use may be without your approval. Most hams remove the microphone to protect a station from being operated without permission. (Before you can go on the air again, you must remember where you hid the mike.)

2D-1.2 What is the purpose of a key-operated "ON/OFF" switch in the main power line?

A. To prevent the use of your station by unauthorized persons

B. To provide an easy method for the FCC to put your station off the air

C. To prevent the power company from inadvertently turning off your electricity during an emergency

D. As a safety feature, to kill all power to the station in the event of an emergency

ANSWER A: While I have never seen a ham station with a key-operated "on/off" switch, it is still a good idea to consider some way of securely turning off your set with a secret switch. This way no one will accidentally energize it and use it.

2D-2.1 Why should all antenna and rotator cables be grounded when an amateur station is not in use?

A. To lock the antenna system in one position

B. To avoid radio frequency interference

C. To save electricity

D. To protect the station and building from damage due to a nearby lightning strike

ANSWER D: An easy way to provide an automatic antenna grounding circuit is to use one of those new multi-position antenna selector switches. Purposely ground one of the connections so that you can switch to it when your equipment is not in use.

2D-2.2 How can an antenna system be protected from damage caused by a nearby lightning strike?

A. Install a balun at the antenna feed point.

B. Install an RF choke in the feed line.

C. Ground all antennas when they are not in use.

D. Install a line fuse in the antenna wire.

ANSWER C: Always use a good quality amateur-grade grounded antenna selector switch for lightning protection. You also can install lightning arrestors.

2D-2.3 How can amateur station equipment be protected from damage caused by voltage induced in the power lines by a nearby lightning strike?

A. Use heavy insulation on the wiring.

B. Keep the equipment on constantly.

C. Disconnect the ground system.

D. Disconnect all equipment after use, either by unplugging or by using a main disconnect switch.

ANSWER D: If you suspect a lightning storm approaching, unplug everything, and actually remove the set from other grounded equipment in your ham shack.

2D-2.4 For proper protection from lightning strikes, what equipment should be grounded in an amateur station?

A. The power supply primary

B. All station equipment

C. The feed line center conductors

D. The ac power mains

ANSWER B: A good ground system will minimize noise pick-up, reduce radio frequency interference (RFI), enhance your communications range, and provide safety at your ham station. A thick copper braid (such as from discarded RG-8 coaxial cable) is inexpensive and makes an excellent ground. Copper foil is the best.

2D-3.1 What is a convenient indoor grounding point for an amateur station?
A. A metallic cold water pipe
B. PVC plumbing
C. A window screen
D. A natural gas pipe

ANSWER A: Run your copper foil from the chassis of each piece of equipment to a common grounding bar (a length of 2-inch diameter copper water pipe is ideal). Connect the grounding bar to a good earth ground, either a bar driven into the ground for at least two feet, or a cold water pipe that goes underground. Avoid hot water pipes since they do not run directly to earth. Don't use a cold water pipe grounding system if you have PVC plastic water pipes! Instead, use a 6- to 8-foot copper-plated steel ground rod that's available from Radio Shack.

2D-3.2 To protect against electrical shock hazards, what should you connect the chassis of each piece of your equipment to?
A. Insulated shock mounts
B. The antenna
C. A good ground connection
D. A circuit breaker

ANSWER C: When you add new gear to your ham station set-up, always connect the chassis to the other gear with a good common braid or foil connection. Be certain all connections are tight! You'll have less ground noise that can affect transmitted and received signals.

2D-3.3 What type of material should a driven ground rod be made of?
A. Ceramic or other good insulator
B. Copper or copper-clad steel
C. Iron or steel
D. Fiberglass

ANSWER B: Be careful where you drive in your copper/clad steel ground rod. One time your author managed to pound it through a gas pipe. What a big surprise!

2D-3.4 What is the shortest ground rod you should consider installing for your amateur station RF ground?
A. 4 foot
B. 6 foot
C. 8 foot
D. 10 foot

ANSWER C: Use a sledge hammer to get the ground rod firmly into the soil. Always wear protective goggles when hammering away at the rod. Gloves are a good idea to prevent blisters, too. Quote: "This is the voice of experience!"

One question must be from the following 13 questions:

2D-4.1 What precautions should you take when working with 1270-MHz waveguide?
A. Make sure that the RF leakage filters are installed at both ends of the waveguide.

B. Never look into the open end of a waveguide when RF is applied.

C. Minimize the standing wave ratio before you test the waveguide.

D. Never have both ends of the waveguide open at the same time when RF is applied.

ANSWER B: A waveguide is a form of low-loss transmission line that transports microwave energy from one place to another much like a garden hose carries water. The RF radiation bounces off of the sides of the waveguide along its route. Concentrated microwave radiation is generally considered to be biologically harmful to humans—particularly to the eyes, so don't needlessly expose yourself to this hazard. Be sure to keep your face away from the antenna or waveguide at these frequencies.

2D-4.2 What precautions should you take when you mount a UHF antenna in a permanent location?

A. Make sure that no one can be near the antenna when you are transmitting.

B. Make sure that the RF field screens are in place.

C. Make sure that the antenna is near the ground to maximize directional effect.

D. Make sure you connect an RF leakage filter at the antenna feed point.

ANSWER A: Antennas are dangerous when you are transmitting. Even though you may be putting out just a few watts of power, ultra high frequency (UHF) radio waves may be harmful to your body within a few inches of the antenna system. Keep everyone well clear.

2D-4.3 What precautions should you take before removing the shielding on a UHF power amplifier?

A. Make sure all RF screens are in place at the antenna.

B. Make sure the feed line is properly grounded.

C. Make sure the amplifier cannot be accidentally energized.

D. Make sure that the RF leakage filters are connected.

ANSWER C: If you are working on anything with high voltage, or potential amounts of UHF output, be cautious. Make absolutely sure no one can accidentally turn on the equipment while your fingers are inside.

2D-4.4 Why should you use only good-quality, well-constructed coaxial cable and connectors for a UHF antenna system?

A. To minimize RF leakage

B. To reduce parasitic oscillations

C. To maximize the directional characteristics of your antenna

D. To maximize the standing wave ratio of the antenna system

ANSWER A: When assembling your VHF and UHF antenna system, use rigid coaxial cable or the highest quality and best coaxial you can purchase at the ham radio store. Never use coaxial cable intended for use at lower frequencies.

2D-4.5 Why should you be careful to position the antenna of your 220-MHz handheld transceiver away from your head when you are transmitting?

A. To take advantage of the directional effect

B. To minimize RF exposure

C. To use your body to reflect the signal, improving the directional characteristics of the antenna

D. To minimize static discharges

ANSWER B: Since they have attached antennas, handheld VHF/UHF ham transceivers expose users to the most radio frequency energy. Aim the antenna away from your head and eyes as much as possible!

2D-4.6 Which of the following types of radiation produce health risks most like the risks produced by radio frequency radiation?
A. Microwave oven radiation and ultraviolet radiation
B. Microwave oven radiation and radiation from an electric space heater
C. Radiation from Uranium or Radium and ultraviolet radiation
D. Sunlight and radiation from an electric space heater

ANSWER B: This is a tricky one. They are not asking what produces microwave radiation. They are asking what device is similar to radio frequency output. Believe it or not, the electric space heater produces a low-level RF type of output.

2D-5.1 Why is there a switch that turns off the power to a high-voltage power supply if the cabinet is opened?
A. To prevent RF from escaping from the supply
B. To prevent RF from entering the supply through the open cabinet
C. To provide a way to turn the power supply on and off
D. To reduce the danger of electrical shock

ANSWER D: Most linear amplifiers incorporate an automatic disconnect switch when you lift the cover. This usually shorts out the high voltage that may still remain in big capacitors. Never circumvent this switch—it's there for your safety.

2D-5.2 What purpose does a safety interlock on an amateur transmitter serve?
A. It reduces the danger that the operator will come in contact with dangerous high voltages when the cabinet is opened while the power is on.
B. It prevents the transmitter from being turned on accidentally.
C. It prevents RF energy from leaking out of the transmitter cabinet.
D. It provides a way for the station licensee to ensure that only authorized operators can turn the transmitter on.

ANSWER A: Never, never, never open the lid of a linear amplifier when the set is turned on. There is a terrific bang when the interlock switch shorts out the high-voltage dc.

2D-6.1 What type of safety equipment should you wear when you are working at the top of an antenna tower?
A. A grounding chain
B. A reflective vest
C. Loose clothing
D. A carefully inspected safety belt

ANSWER D: Not all safety belts are safe. The old leather lineman's belt is probably brittle and may break when you are up on the tower. Always check out your climbing equipment well before going aloft.

2D-6.2 Why should you wear a safety belt when you are working at the top of an antenna tower?
A. To provide a way to safely hold your tools so they don't fall and injure someone on the ground

B. To maintain a balanced load on the tower while you are working
C. To provide a way to safely bring tools up and down the tower
D. To prevent an accidental fall

ANSWER D: Your author has personally slipped on a rung of a tower, and the safety belt saved his life. It could save your life, too.

2D-6.3 For safety purposes, how high should you locate all portions of your horizontal wire antenna?

A. High enough so that a person cannot touch them from the ground
B. Higher than chest level
C. Above knee level
D. Above electrical lines

ANSWER A: You don't need much height for a simple 10-meter dipole. However, keep it at least 15 feet off the ground. This gives you better range, and it also prevents anyone from accidentally touching it.

2D-6.4 What type of safety equipment should you wear when you are on the ground assisting someone who is working on an antenna tower?

A. A reflective vest
B. A safety belt
C. A grounding chain
D. A hard hat

ANSWER D: Local hams may invite you to an "antenna party." As a group, you may help another ham put up the antenna atop a tower. If you are on the ground, make sure you wear a hard hat and safety glasses.

2D-6.5 Why should you wear a hard hat when you are on the ground assisting someone who is working on an antenna tower?

A. To avoid injury from tools dropped from the tower
B. To provide an RF shield during antenna testing
C. To avoid injury if the tower should accidentally collapse
D. To avoid injury from walking into tower guy wires

ANSWER A: Quote: "If you are assisting *ME* in a tower project, I'll be dropping tools all over the place, so be sure to wear that safety hat and protective glasses."

One question must be from the following 10 questions:

2D-7-1.1 What accessory is used to measure standing wave ratio?

A. An ohm meter
B. An ammeter
C. An SWR meter
D. A current bridge

ANSWER C: *Standing wave ratio* is the comparison of the power going forward from the transmitter to the power reflected back from the load—usually an antenna. It is measured by a standing wave ratio bridge—sometimes called a *reflectometer*, since it measures reflected power. SWR bridges for use below 30 MHz can be purchased for as little as $30. For VHF/UHF frequencies, a more expensive (about $150) bridge must be used for maximum accuracy.

2D-7-1.2 What instrument is used to indicate the relative impedance match between a transmitter and antenna?
A. An ammeter
B. An ohmmeter
C. A voltmeter
D. An SWR meter

ANSWER D: Impedance is similar to the back pressure that a tuned muffler presents to an engine in a sports car. We measure the impedance match between the "back pressure" of your transmitter and antenna with a standing wave ratio (SWR) meter.

2D-7-2.1 What does an SWR meter reading of 1:1 indicate?
A. An antenna designed for use on another frequency band is probably connected
B. An optimum impedance match has been attained
C. No power is being transferred to the antenna
D. An SWR meter never indicates 1:1 unless it is defective

ANSWER B: A 1:1 SWR is a perfect match. However, it doesn't necessarily mean you're going to have a great signal on the airwaves. You can achieve a 1:1 SWR by transmitting into a non-radiating dummy load!

2D-7-2.2 What does an SWR meter reading of less than 1.5:1 indicate?
A. An unacceptably low reading
B. An unacceptably high reading
C. An acceptable impedance match
D. An antenna gain of 1.5

ANSWER C: If you have a 1.5:1 SWR, not bad! But again, this reading is meaningless as a guarantee of good range.

2D-7-2.3 What does an SWR meter reading of 4:1 indicate?
A. An unacceptably low reading
B. An acceptable impedance match
C. An antenna gain of 4
D. An impedance mismatch, which is not acceptable; it indicates problems with the antenna system

ANSWER D: A 4:1 is a sure indication that something is wrong with your antenna system. Here the SWR surely indicates you won't get many contacts. See what's wrong.

2D-7-2.4 What does an SWR meter reading of 5:1 indicate?
A. The antenna will make a 10-watt signal as strong as a 50-watt signal
B. Maximum power is being delivered to the antenna
C. An unacceptable mismatch is indicated
D. A very desirable impedance match has been attained

ANSWER C: At 5:1, your rig is sure to temporarily shut down on transmit. Probably a shorted coaxial connector. Something is really wrong.

2D-7-3.1 What kind of SWR meter reading may indicate poor electrical contact between parts of an antenna system?
A. An erratic reading
B. An unusually low reading

C. No reading at all
D. A negative reading

ANSWER A: It's a good idea to watch your SWR meter while transmitting—especially if the wind is strong. If the wind is blowing your antenna around, you may notice erratic or intermittent readings on your SWR bridge. Guess what? You have a bad connection aloft.

2D-7-3.2 What does an unusually high SWR meter reading indicate?

A. That the antenna is not the correct length, or that there is an open or shorted connection somewhere in the feed line
B. That the signals arriving at the antenna are unusually strong, indicating good radio conditions
C. That the transmitter is producing more power than normal, probably indicating that the final amplifier tubes or transistors are about to go bad
D. That there is an unusually large amount of solar white-noise radiation, indicating very poor radio conditions

ANSWER A: Another cause of high SWR, not mentioned in the answer, is your antenna is too close to the earth. The 10-meter dipole for your new Novice voice privileges should be at least 15 feet above the ground. It won't work on the ground.

2D-7-3.3 The SWR meter reading at the low-frequency end of an amateur band is 2.5:1, and the SWR meter reading at the high-frequency end of the same band is 5:1. What does this indicate about your antenna?

A. The antenna is broadbanded.
B. The antenna is too long for operation on this band.
C. The antenna is too short for operation on this band.
D. The antenna has been optimized for operation on this band.

ANSWER B: If you go slightly higher in frequency, and the SWR continues to climb, your antenna is too long. Cut off an inch and see what happens.

2D-7-3.4 The SWR meter reading at the low-frequency end of an amateur band is 5:1, and the SWR meter reading at the high-frequency end of the same band is 2.5:1. What does this indicate about your antenna?

A. The antenna is broadbanded.
B. The antenna is too long for operation on this band.
C. The antenna is too short for operation on this band.
D. The antenna has been optimized for operation on this band.

ANSWER C: Here is just the opposite situation—as you tune higher in frequency within your band limits, the SWR gets better and better. Lengthen the antenna slightly for a better match. If you want to continue to operate at the low end of the band, lengthen the antenna more. Remember "lower-longer" as you tune to get better SWR at the low end of the band.

One question must be from the following 12 questions:

2D-8-1.1 What is meant by receiver overload?

A. Interference caused by transmitter harmonics
B. Interference caused by overcrowded band conditions
C. Interference caused by strong signals from a nearby transmitter
D. Interference caused by turning the receiver volume too high

ANSWER C: Many times a high-pass filter installed at the antenna input of a television will help reduce television interference (TVI). It is the responsibility of the set owner—not the ham operator—to install this filter.

2D-8-1.2 What is a likely indication that radio-frequency interference to a receiver is caused by front-end overload?
- A. A low-pass filter at the transmitter reduces interference sharply.
- B. The interference is independent of frequency.
- C. A high-pass filter at the receiver reduces interference little or not at all.
- D. Grounding the receiver makes the problem worse.

ANSWER B: If you mount your transmitting antenna too close to other receiving antennas used for home entertainment equipment, chances are your transmitted signal will come in on your radio or television receiver. If the interference is independent of your transmitter's location, or where the receiver or TV is tuned, the problem is proximity of the two antennas. This is called "front-end overload."

If you see a lot of outside television antennas on roofs around you, chances are your next door neighbors will need to use a *high-pass filter* on their TV set to prevent interference from your signal. High-pass filters reject all frequencies below a cutoff frequency. The cutoff point (usually around 45 MHz) is higher than your 10-meter transmitted ham signals, but lower than the 55.25 MHz (channel 2) of the first VHF television channel.

2D-8-1.3 Your neighbor reports interference to his television whenever you are transmitting from your amateur station. This interference occurs regardless of your transmitter frequency. What is likely to be the cause of the interference?
- A. Inadequate transmitter harmonic suppression
- B. Receiver VR tube discharge
- C. Receiver overload
- D. Incorrect antenna length

ANSWER C: Receiver overload may be reduced by keeping your transmitting antenna as far away from other antennas as possible. Reducing power will also help. Good grounding techniques will help. Staying off the air during big ball games may also help!

2D-8-1.4 What type of filter should be installed on a TV receiver as the first step in preventing RF overload from an amateur HF station transmission?
- A. Low pass
- B. High pass
- C. Band pass
- D. Notch

ANSWER B: The TV band is higher than your worldwide radio and your new voice privileges on 10 meters. Putting a *high-pass filter* on the TV receiver, if it's connected to an outside antenna, is a good start in cleaning up interference. Don't put a high-pass filter into a cable TV system. Cable television feeds will not work through a high-pass filter.

2D-8-2.1 What is meant by *harmonic radiation*?
- A. Transmission of signals at whole number multiples of the fundamental (desired) frequency
- B. Transmission of signals that include a superimposed 60-Hz hum

C. Transmission of signals caused by sympathetic vibrations from a nearby transmitter

D. Transmission of signals to produce a stimulated emission in the air to enhance skip propagation

ANSWER A: Every transmitted signal contains a weak second harmonic. If you operate at 7.1 MHz, your second harmonic will be 14.2 MHz, which is in another ham band. If you operate at 28.2 MHz, your second harmonic will land at 56.4 MHz, which is in the middle of television channel 2. Be aware of where your second harmonic falls to be a good operator and a good neighbor!

2D-8-2.2 Why is harmonic radiation from an amateur station undesirable?

A. It will cause interference to other stations and may result in out-of-band signal radiation.

B. It uses large amounts of electric power.

C. It will cause sympathetic vibrations in nearby transmitters.

D. It will produce stimulated emission in the air above the transmitter, thus causing aurora.

ANSWER A: A modern ham radio equipment for worldwide operation will contain harmonic suppressors, and so do power amplifiers. Sometimes an *external low-pass filter* for your high frequency radio will reduce the second harmonic. Low-pass filters have little insertion signal loss up to a certain cutoff frequency above your 10 meter transmitted signal. They will reject the higher harmonic or un-wanted extraneous frequencies. One of the best ways to reduce harmonic interference, however, is to simply reduce your transmitter power level. Don't run more power than you need. There is no point in running a linear amplifier if a "barefoot" 100 watter does just as well—*and most will!* You will be surprised how little signal gain there is between 100 and 1000 watts! There could be a very big difference in TVI levels, however.

2D-8-2.3 What type of interference may radiate from a multi-band antenna connected to an improperly tuned transmitter?

A. Harmonic radiation

B. Auroral distortion

C. Parasitic excitation

D. Intermodulation

ANSWER A: Today's modern transistorized transceivers do not require power-amplifier tuning; however, older rigs that use tubes in the final stage require tuning. Improperly tuned transmitters which feed into multi-band antennas can result in harmonic radiation. Tube rigs are fine, but the new solid-state transistor-ized sets are best, and you won't have to worry about proper tune-up proce-dures.

2D-8-2.4 What is the purpose of shielding in a transmitter?

A. It gives the low pass filter structural stability.

B. It enhances the microphonic tendencies of radiotelephone transmitters.

C. It prevents unwanted RF radiation.

D. It helps maintain a sufficiently high operating temperature in circuit components.

ANSWER C: Never operate your equipment with the metal cabinet removed. If you are working on your equipment, always reassemble it completely before going on the air. Transmitting with the covers off could lead to unwanted RF radiation escaping from circuits that should be enclosed.

2D-8-2.5 Your neighbor reports interference on one or two channels of her television when you are transmitting from your amateur station. This interference only occurs when you are operating on 15 meters. What is likely to be the cause of the interference?
A. Excessive low-pass filtering on the transmitter
B. Sporadic E de-ionization near your neighbor's TV antenna
C. TV Receiver front-end overload
D. Harmonic radiation from your transmitter
ANSWER D: If your neighbor is using an outside TV antenna, they might experience multiples of your transmitted signal on the 15-meter and 10-meter bands. These multiples, called harmonics, are minimized by a low-pass filter on your worldwide ham set. Another cure is to encourage your neighbors to sign up for cable television which is less susceptible to interference.

2D-8-2.6 What type of filter should be installed on an amateur transmitter as the first step in reducing harmonic radiation?
A. Key click filter
B. Low pass filter
C. High pass filter
D. CW filter
ANSWER B: Low-pass filters are designed only for 10-80-meter transceivers (the worldwide sets used on 10 meters). Never put a low-pass filter on your 220-MHz or 1270-MHz sets because these bands are much higher than low-pass frequencies. Your VHF and UHF FM equipment seldom interferes with TVs or hi-fi sets.

2D-8-3.1 If you are notified that your amateur station is causing television interference, what should you do first?
A. Make sure that your amateur equipment is operating properly, and that it does not cause interference to your own television.
B. Immediately turn off your transmitter and contact the nearest FCC office for assistance.
C. Install a high-pass filter at the transmitter output and a low-pass filter at the antenna-input terminals of the TV.
D. Continue operating normally, since you have no legal obligation to reduce or eliminate the interference.
ANSWER A: On 10 meters with your new voice privileges, chances are you may generate some television interference. This is minimized by *low-pass filters* on your ham set, *high-pass filters* on your neighbor's TV antenna and receiver, or a good cable company feed system. Start out with your own television—see if you are affecting it.

2D-8-3.2 Your neighbor informs you that you are causing television interference, but you are sure your amateur equipment is operating properly and you cause no interference to your own TV. What should you do?
A. Immediately turn off your transmitter and contact the nearest FCC office for assistance.
B. Work with your neighbor to determine that you are actually the cause of the interference.

C. Install a high-pass filter at the transmitter output and a low-pass filter at the antenna-input terminals of the TV.

D. Continue operating normally, since you have no legal obligation to reduce or eliminate the interference.

ANSWER B: An irate neighbor will cause you no end of headaches until you quit interfering with their TV set. Work with them to illustrate you are a good, concerned ham and are truly interested in resolving the problem. Don't ignore them—the problem won't go away naturally (unless they move or subscribe to cable television).

Subelement 2E – Electrical Principles (4 examination questions from 44 questions in 2E)

One question must be from the following 10 questions:

2E-1-1.1 Your receiver dial is calibrated in megahertz and shows a signal at 1200 MHz. At what frequency would a dial calibrated in gigahertz show the signal?

A. 1,200,000 GHz
B. 12 GHz
C. 1.2 GHz
D. 0.0012 GHz

ANSWER C: One gigahertz (1×10^9) is equal to 1,000 megahertz ($1,000 \times 1 \times 10^6$). To convert megahertz to gigahertz, move the decimal point 3 places to the *left*—1200 MHz is 1.2 GHz.

2E-1-2.1 Your receiver dial is calibrated in kilohertz and shows a signal at 7125 kHz. At what frequency would a dial calibrated in megahertz show the signal?

A. 0.007125 MHz
B. 7.125 MHz
C. 71.25 MHz
D. 7,125,000 MHz

ANSWER B: One megahertz (1×10^6) is equal to 1,000 kilohertz ($1000 \times 1 \times 10^3$). To convert kHz to MHz, move the decimal point 3 places to the *left*—7125 kHz is 7.125 MHz. If you want to convert MHz to meters; (i.e., find the wavelength of a frequency), divide MHz into 300.

2E-1-2.2 Your receiver dial is calibrated in gigahertz and shows a signal at 1.2 GHz. At what frequency would a dial calibrated in megahertz show the same signal?

A. 1.2 MHz
B. 12 MHz
C. 120 MHz
D. 1200 MHz

ANSWER D: Your author has never seen a radio calibrated in GHz. But if it is, move the decimal point 3 places to the *right* to convert GHz to MHz. 1.2 GHz is 1200 MHz. This is the opposite of Question 2E-1-1.1.

2E-1-3.1 Your receiver dial is calibrated in megahertz and shows a signal at 3.525 MHz. At what frequency would a dial calibrated in kilohertz show the signal?
  A. 0.003525 kHz
  B. 3525 kHz
  C. 35.25 kHz
  D. 3,525,000 kHz

ANSWER B: Move the decimal point 3 places to the *right* to convert MHz to kHz. 3.525 MHz is 3525 kHz. This is the opposite of Question 2E-1-2.1.

2E-1-3.2 Your receiver dial is calibrated in kilohertz and shows a signal at 3725 kHz. At what frequency would a dial calibrated in hertz show the same signal?
  A. 3,725 Hz
  B. 3.725 Hz
  C. 37.25 Hz
  D. 3,725,000 Hz

ANSWER D: Here's another silly answer you will probably never encounter out there in the ham radio world. Kilo stands for 1,000 (1×10^3). Thus, 3725 kHz is really $3725 \times 1 \times 10^3$ or 3,725,000 hertz. Hertz (Hz) means cycles per second.

2E-1-4.1 How long (in meters) is an antenna that is 400 centimeters long?
  A. 0.0004 meters
  B. 4 meters
  C. 40 meters
  D. 40,000 meters

ANSWER B: There are 100 centimeters in one meter, so divide centimeters by 100 to convert to meters. Or move the decimal point 2 places to the left. Calculator keystrokes are: CLEAR 400 ÷ 100 = and the answer is 4.

2E-1-5.1 What reading will be displayed on a meter calibrated in amperes when it is being used to measure a 3000-milliampere current?
  A. 0.003 amperes
  B. 0.3 amperes
  C. 3 amperes
  D. 3,000,000 amperes

ANSWER C: One milliampere equals one one-thousandth of an ampere (1×10^{-3}); therefore, one ampere equal 1000 milliamperes. Divide milliamperes by 1,000 to convert to amperes. Or move the decimal point 3 places to the left. Calculator keystrokes are: CLEAR 3000 ÷ 1000 = and the answer is 3.

2E-1-5.2 What reading will be displayed on a meter calibrated in volts when it is being used to measure a 3500-millivolt potential?
  A. 350 volts
  B. 35 volts
  C. 3.5 volts
  D. 0.35 volts

ANSWER C: Since milli means 1/1,000, one volt equals 1000 millivots. Move the decimal point 3 places to the left to convert millivolts to volts. This is the same as dividing millivolts by 1000. Calculator keystrokes are: CLEAR 3500 ÷ 1000 = and the answer is 3.5.

2E-1-6.1 How many farads is 500,000 microfarads?
  A. 0.0005 farads
  B. 0.5 farads
  C. 500 farads
  D. 500,000,000 farads
ANSWER B: A microfarad is one millionth (1×10^{-6}) of a farad. One farad equals one million microfarads. Move the decimal point 6 places to the left to convert from microfarads to farads. This is the same as dividing microfarads by 1,000,000. Calculator keystrokes are: CLEAR 500000 ÷ 1000000 = and the answer is .5.

2E-1-7.1 How many microfarads is 1,000,000 picofarads?
  A. 0.001 microfarads
  B. 1 microfarad
  C. 1,000 microfarads
  D. 1,000,000,000 microfarads
ANSWER B: A picofarad is one millionth of a microfarad or one million millionth of a farad. Move the decimal point 6 places to the left to convert to microfarads or 12 places to the left to convert to farads. 1,000,000 picofarads ÷ 1,000,000 = 1 microfarad and 1 microfarad ÷ 1,000,000 = 0.000001 farad. Each place that the decimal point is moved to the left is equivalent to dividing by 10.

One question must be from the following 11 questions:

2E-2-1.1 What is the term used to describe the flow of electrons in an electric circuit?
  A. Voltage
  B. Resistance
  C. Capacitance
  D. Current
ANSWER D: Think of the flow of electrons as the flow of water in a stream. If you get out there in midstream, you will feel the current.

2E-2-2.1 What is the basic unit of electric current?
  A. The volt
  B. The watt
  C. The ampere
  D. The ohm
ANSWER C: The flow of electrons in a conductor is called *current*. Current is measured in amperes. Amperes is often referred to as "amps."

2E-3-1.1 What supplies the force that will cause electrons to flow through a circuit?
  A. Electromotive force, or voltage
  B. Magnetomotive force, or inductance
  C. Farad force, or capacitance
  D. Thermodynamic force, or entropy
ANSWER A: Electronic circuits contain electromotive force, or voltage. This causes electrons to flow.

2E-3-1.2 The pressure in a water pipe is comparable to what force in an electrical circuit?

A. Current
B. Resistance
C. Gravitation
D. Voltage

ANSWER D: Until you open the flood gate, you have no water flow. Until you turn the switch on, the pressure is just waiting, ready to do something.

2E-3-1.3 An electric circuit must connect to two terminals of a voltage source. What are these two terminals called?

A. The north and south poles
B. The positive and neutral terminals
C. The positive and negative terminals
D. The entrance and exit terminals

ANSWER C: Any dc source, such as a battery, has a positive and negative terminal.

2E-3-2.1 What is the basic unit of voltage?

A. The volt
B. The watt
C. The ampere
D. The ohm

ANSWER A: Volts are volts, the same as voltage!

2E-4.1 List at least three good electrical conductors.

A. Copper, gold, mica
B. Gold, silver, wood
C. Gold, silver, aluminum
D. Copper, aluminum, paper

ANSWER C: Most wire is copper, and this is a good conductor. Some relays use gold- or silver-plated contacts, and these are also good conductors. You can use aluminum foil as a ground plane; it also is a good conductor. Always read all answers completely—mica, wood and paper are insulators!

2E-5.1 List at least four good electrical insulators.

A. Glass, air, plastic, porcelain
B. Glass, wood, copper, porcelain
C. Paper, glass, air, aluminum
D. Plastic, rubber, wood, carbon

ANSWER A: In order for there to be current from one point to another in a circuit, current must have a good path of conductivity. If there is a poor or open connection, there will be little or no current.

2E-6-1.1 There is a limit to the electric current that can pass through any material. What is this current limiting called?

A. Fusing
B. Reactance
C. Saturation
D. Resistance

ANSWER D: In a river, there is a limit as to how much current will flow down stream. Logs and rocks offer "resistance" to the river current. Similarly, certain materials or small diameter wires in an electric circuit offer resistance to electric current.

2E-6-1.2 What is an electrical component called that opposes electron movement through a circuit?
A. A resistor
B. A reactor
C. A fuse
D. An oersted

ANSWER A: Resistors resist movement of electrons through a circuit.

2E-6-2.1 What is the basic unit of resistance?
A. The volt
B. The watt
C. The ampere
D. The ohm

ANSWER D: The basic unit of resistance is called the ohm.

One question must be from the following 11 questions:

2E-7.1 What electrical principle relates voltage, current and resistance in an electric circuit?
A. Ampere's Law
B. Kirchhoff's Law
C. Ohm's Law
D. Tesla's Law

ANSWER C: The relationship between voltage, current, and resistance in an electric circuit is called "Ohm's Law."

2E-7.2 There is a 2-amp current through a 50-ohm resistor. What is the applied voltage?
A. 0.04 volts
B. 52 volts
C. 100 volts
D. 200 volts

ANSWER C: A simple way to calculate Ohm's Law is to use the Ohm's Law magic circle:

E = VOLTAGE IN VOLTS
I = CURRENT IN AMPERES
R = RESISTANCE IN OHMS

Ohm's Law Calculation

Ohm's Law ($E = I \times R$) states a relationship between voltage, current and resistance in an electrical circuit. It says that the applied electromotive force, E, in volts, is equal to the circuit current, I, in amperes, times the circuit resistance, R, in ohms. You can solve for E, I or R if the other two quantities are known. The three equations are:

$$E = I \times R \qquad\qquad I = \frac{E}{R} \qquad\qquad R = \frac{E}{I}$$

The "magic circle" helps you remember these equations. To use it, cover the unknown quantity with your finger and solve the equation for the remaining quantities. If you know the values of I and R and want to find the value of E, cover the E in the magic circle and it shows that you must multiply I times R. If you want to find I, cover the I and it shows that you must divide E by R. If you want to find R, cover the R and it shows that you must divide E by I. Since we are looking for E, the applied voltage, cover E with your finger, and you now have I (2 amps) times R (50 ohms). Multiply these two to obtain your answer of 100 volts. The calculator keystrokes are: CLEAR 2 × 50 =.

2E-7.3 If 200 volts is applied to a 100-ohm resistor, what is the current through the resistor?
   A. 0.5 amps
   B. 2 amps
   C. 50 amps
   D. 20000 amps
ANSWER B: Refer to Question 2E-7.2. In this problem, you are looking for R. Using the Ohm's Law magic circle, cover R with your finger. You now have E over I, or 200 over 100. Do the division, and you will end up with 2 amps. See how simple this is! Calculator keystrokes are: CLEAR 200 ÷ 100 =.

2E-7.4 There is a 3-amp current through a resistor and we know that the applied voltage is 90 volts. What is the value of the resistor?
   A. 0.03 ohms
   B. 10 ohms
   C. 30 ohms
   D. 2700 ohms
ANSWER C: Refer to Question 2E-7.2. In this problem, you want to find R. Covering R with your finger leaves E over I. 90 divided by 3 gives 30 ohms. See how simple it is to use Ohm's Law. Calculator keystrokes are: CLEAR 90 ÷ 3 = .

2E-8.1 What is the term used to describe the ability to do work?
   A. Voltage
   B. Power
   C. Inertia
   D. Energy
ANSWER D: Energy is the ability to do work. When you work hard reviewing these questions, you are burning up energy. Having plenty of energy will help you pass the test.

2E-8.2 What is converted to heat and light in an electric light bulb?
   A. Electrical energy
   B. Electrical voltage
   C. Electrical power
   D. Electrical current
ANSWER A: Turn the light on, and it consumes energy. Look at that watt-hour meter on the side of your house—see how fast it turns when you turn everything on!

2E-9-1.1 What term is used to describe the rate of energy consumption?
   A. Energy
   B. Current
   C. Power
   D. Voltage

ANSWER C: Power indicates the rate of energy being consumed.

2E-9-1.2 You have two lamps with different wattage light bulbs in them. How can you determine which bulb uses electrical energy faster?
    A. The bulb that operates from the higher voltage will consume energy faster
    B. The physically larger bulb will consume energy faster
    C. The bulb with the higher wattage rating will consume energy faster
    D. The bulb with the lower wattage rating will consume energy faster

ANSWER C: Light bulbs that glow with different intensities consume different amounts of energy. Usually, the higher wattage bulb consumes energy faster, that is, it consumes more power.

2E-9-2.1 What is the basic unit of electrical power?
    A. Ohm
    B. Watt
    C. Volt
    D. Ampere

ANSWER B: Power is energy, and you all have one of those energy meters on the side of your house. You know, that's the meter that keeps turning after you've turned just about everything off! Volts times amps equals watts. There is a "magic circle" for power calculation that is similar to the one for Ohm's Law. Here it is:

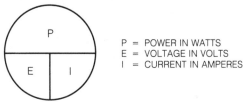

P = POWER IN WATTS
E = VOLTAGE IN VOLTS
I = CURRENT IN AMPERES

Power Calculation

As shown, P = power in watts, E = voltage in volts, and I = current in amperes. Use it in the same way as you use the Ohm's Law magic circle; that is, cover the unknown quantity with your finger and perform the mathematical operation represented by the remaining quantities. (Refer to Question 2E-7.2.)

2E-10.1 What is the term for an electrical circuit in which there can be no current?
    A. A closed circuit
    B. A short circuit
    C. An open circuit
    D. A hyper circuit

ANSWER C: If you turn on your equipment and nothing happens at all, probably something is "open," like that hidden power switch.

2E-11.1 What is the term for a failure in an electrical circuit that causes excessively high current?
    A. An open circuit
    B. A dead circuit
    C. A closed circuit
    D. A short circuit

ANSWER D: Anytime you have a malfunction of a piece of equipment, and you hear a pop or smell something burning, chances are a short circuit has caused the malfunction.

One question must be from the following 12 questions:

2E-12-1.1 What is the term used to describe a current that flows only in one direction?
  A. Alternating current
  B. Direct current
  C. Periodic current
  D. Pulsating current
ANSWER B: Batteries generate direct current. Even though a current may vary in value, if it always flows in the same direction, it is a direct current (dc).

2E-12-2.1 What is the term used to describe a current that flows first in one direction, then in the opposite direction, over and over?
  A. Alternating current
  B. Direct current
  C. Negative current
  D. Positive current
ANSWER A: If you have been shocked by house power, chances are you felt the "buzz." Be careful, it is very dangerous! Direct current (dc) flows in one direction; alternating current (ac) changes direction.

2E-12-3.1 What is the term for the number of complete cycles of an alternating waveform that occur in one second?
  A. Pulse repetition rate
  B. Hertz
  C. Frequency per wavelength
  D. Frequency
ANSWER D: Frequency is measured in cycles per second, called hertz. Counting the number of alternating cycles in one second gives the frequency in hertz.

2E-12-3.2 A certain ac signal makes 2000 complete cycles in one second. What property of the signal does this number describe?
  A. The frequency of the signal
  B. The pulse repetition rate of the signal
  C. The wavelength of the signal
  D. The hertz per second of the signal
ANSWER A: Remember hertz, abbreviated Hz, means cycles per second—and this is frequency. This is a bit tricky, so be careful. Answer D actually says, "The cycles per second per second of the signal."

2E-12-3.3 What is the basic unit of frequency?
  A. The hertz
  B. The cycle
  C. The kilohertz
  D. The megahertz
ANSWER A: The unit of frequency is named after a man named Hertz who developed horizontal antennas. Old timers remember when the unit was cycles per second (cps).

2E-12-4.1 What range of frequencies are usually called audio frequencies?
A. 0 to 20 Hz
B. 20 to 20,000 Hz
C. 200 to 200,000 Hz
D. 10,000 to 30,000 Hz
ANSWER B: Quote: "I'm not sure that I can still hear all the way up to 20,000 hertz (20 kHz). However, my dogs and cats can!"

2E-12-4.2 A signal at 725 Hz is in what frequency range?
A. Audio frequency
B. Intermediate frequency
C. Microwave frequency
D. Radio Frequency
ANSWER A: This frequency is between 20 Hz and 20 kHz, so it is audio.

2E-12-4.3 Why do we call signals in the range 20 Hz to 20,000 Hz audio frequencies?
A. Because the human ear rejects signals in this frequency range
B. Because the human ear responds to sounds in this frequency range
C. Because frequencies in this range are too low for a radio to detect
D. Because a radio converts signals in this range directly to sounds the human ear responds to
ANSWER B: Audio frequencies are those that you hear with your ear. Among the radio frequencies, above 20,000 hertz, are those that you can pick up with your ham receiver.

2E-12-5.1 Signals above what frequency are usually called radio-frequency signals?
A. 20 Hz
B. 2000 Hz
C. 20,000 Hz
D. 1,000,000 Hz
ANSWER C: Radio frequencies are above 20 kHz and audio frequencies are below 20 kHz. The letter "k" represents 1,000 or 3 zeros. It is used as a shorthand so you don't have to write down all those zeros.

2E-12-5.2 A signal at 7125 kHz is in what frequency range?
A. Audio frequency
B. Radio frequency
C. Hyper-frequency
D. Super-high frequency
ANSWER B: Is this above 20,000 hertz? You bet it is, so it must be radio frequency.

2E-13.1 What is the term for the distance an ac signal travels during one complete cycle?
A. Wave velocity
B. Velocity factor
C. Wavelength
D. Wavelength per meter
ANSWER C: A simple way to approximate a certain spot on the radio dial is to use wavelength. For instance, long-range Novice voice privileges are on the 10-

meter band. 10 divided into 300 is 30 MHz (See Question 2A-10.7.) This approximates where you would find your Novice sub-band on 10 meters. The middle of the Novice sub-band on 10 meters is exactly 28.4 MHz, and this is frequency. Frequency is an exact spot, but the nominal wavelength for a particular band is a relatively broad area on the radio dial. Recall that:

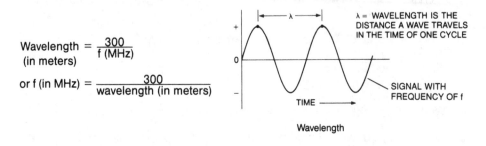

Wavelength $= \dfrac{300}{f \text{ (MHz)}}$
(in meters)

or f (in MHz) $= \dfrac{300}{\text{wavelength (in meters)}}$

λ = WAVELENGTH IS THE DISTANCE A WAVE TRAVELS IN THE TIME OF ONE CYCLE

SIGNAL WITH FREQUENCY OF f

TIME

Wavelength

2E-13.2 In the time it takes a certain radio signal to pass your antenna, the leading edge of the wave travels 12 meters. What property of the signal does this number refer to?
 A. The signal frequency
 B. The wave velocity
 C. The velocity factor
 D. The signal wavelength
ANSWER D: When we refer to the length of a radio wave, we are referring to its wavelength—not its frequency.

Subelement 2F – Circuit Components (2 examination questions from 21 questions in 2F)

One question must be from the following 11 questions:

2F-1.1 What is the symbol used on schematic diagrams to represent a resistor?

 A.   B.   C.   D.

ANSWER B: If this were a stream of water, all those bends would present resistance to the current; likewise, a resistor presents resistance to an electric current.

2F-1.2 What is the symbol used on schematic diagrams to represent a variable resistor or potentiometer?

 A.   B.   C.   D.

ANSWER C: You can change the potential coupled to another part of a circuit with a potentiometer. The arrow through the resistor symbol indicates it represents a variable resistance.

2F-1.3 In Diagram 2F-1, which component is a resistor?
A. 1
B. 2
C. 3
D. 4

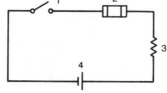

Diagram 2F-1

ANSWER C: Look at the right side of the circuit, and there is the resistor standing straight up and down.

2F-2.1 What is the symbol used on schematic diagrams to represent a single-pole, single-throw switch?

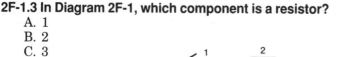

ANSWER A: This is an on/off switch. In one position, it opens a circuit. In the other position, it closes or completes a circuit.

2F-2.2 What is the symbol used on schematic diagrams to represent a single-pole, double-throw switch?

A.   B.   C.   D.

ANSWER A: If you had a double antenna system, such as a 10 meter dipole, and a 10 meter vertical, you would use a single pole, double-throw switch to allow a single radio to work off your double antenna system.

2F-2.3 What is the symbol used on schematic diagrams to represent a double-pole, double-throw switch?

A.   B.   C.   D.

ANSWER B: With this switch, you may throw two contacts in one direction, or in another direction.

2F-2.4 What is the symbol used on schematic diagrams to represent a single-pole, 5-position switch?

ANSWER D: On the schematic of your radio, rotary switches look like Answer D. The rotary switch might allow you to tune in 80 meters, 40 meters, 20 meters, 15 meters, and 10 meters. Five bands requires a 5-pole switch.

2F-2.5 In Diagram 2F-2, which component is a switch?

A. 1
B. 2
C. 3
D. 4

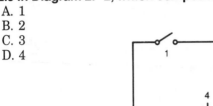

Diagram 2F-2

ANSWER A: That's an easy one to identify—a single-pole single-throw switch. This is an on/off switch.

2F-3.1 What is the symbol used on schematic diagrams to represent a fuse?

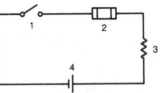

A.　　　　B.　　　　C.　　　　D.

ANSWER C: This little guy is a fuse—an intentional weak link in the circuit. You want this to melt and open the circuit in case of excessive current flow.

2F-4.1 What is the symbol used on schematic diagrams to represent a single-cell battery?

A.　　　　B.　　　　C.　　　　D.

ANSWER C: Battery plates never touch. The symbol looks almost like a capacitor symbol. Notice that the positive side (the long line) is indicated by a plus (+), and the negative side (the short line) is indicated by a minus (−). Watch out for Answer A; the polarity is wrong.

2F-4.2 What is the symbol used on schematic diagrams to represent a multiple-cell battery?

A.　　B.　　C.　　D.

ANSWER B: Here we see additional plates in the battery. Same as 2F-4.1, but more of them. Watch out for incorrect polarity as in Answer D.

One question must be from the following 10 questions:

2F-5.1 What is the symbol normally used to represent an earth-ground connection on schematic diagrams?

A.　　　　B.　　　　C.　　　　D.

ANSWER D: This is the classic ground symbol on worldwide radios and antenna systems. Grounding your equipment and half your antenna system is very important for safety and good range.

2F-5.2 What is the symbol normally used to represent a chassis-ground connection on schematic diagrams?

A. $\wedge$ B. $\overline{}\llap{/\!/\!/}$ C. $\underline{\underline{\perp}}$ D. $\underline{\underline{\wedge}}$

ANSWER B: A chassis ground, rather than an earth soil ground, is indicated by the schematic symbol illustrated in Answer B.

2F-5.3 In Diagram 2F-5, which symbol represents a chassis ground connection?
A. 1
B. 2
C. 3
D. 4

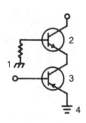

Diagram 2F-5

ANSWER A: You can see the chassis ground, to the left. The earth ground is at the bottom of the diagram.

2F-5.4 In Diagram 2F-5, which symbol represents an earth ground connection?
A. 1
B. 2
C. 3
D. 4

ANSWER D: There's that classic earth ground again.

2F-6.1 What is the symbol used to represent an antenna on schematic diagrams?

A. $\overline{\overline{\top}}$ B. $\Box$ C. Ψ D. Y

ANSWER D: Looks like an antenna, doesn't it? Don't let Answer A or C fool you. You already know what they are.

2F-7.1 What is the symbol used to represent an NPN bipolar transistor on schematic diagrams?

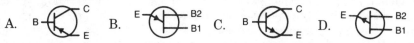

ANSWER C: An easy way to identify an NPN transistor is to first identify base (B), collector (C), and emitter (E). Then look and see which way the arrow is pointing. If the arrow is <u>NOT POINTING IN</u>, then it's an **NPN** transistor.

2F-7.2 What is the symbol used to represent a PNP bipolar transistor on schematic diagrams?

A. ![B-C-E symbol] B. ![E-B2-B1 symbol] C. ![B-C-E symbol] D. ![E-B2-B1 symbol]

ANSWER A: Here the arrow is <u>P</u>OINTING <u>IN</u> so it is a **PNP** transistor.

2F-7.3 In Diagram 2F-7, which symbol represents a PNP bipolar transistor?
- A. 1
- B. 2
- C. 3
- D. 4

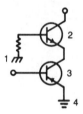

Diagram 2F-7

ANSWER C: Which way is the arrow pointing in the lower transistor? It's <u>P</u>OINT-ING <u>IN</u>, so it's a PNP.

2F-7.4 In Diagram 2F-7, which symbol represents an NPN bipolar transistor?
- A. 1
- B. 2
- C. 3
- D. 4

ANSWER B: In the upper transistor, #2, it's <u>N</u>OT <u>P</u>OINTING <u>IN</u>, so it is an NPN transistor.

2F-8.1 What is the symbol used to represent a triode vacuum tube on schematic diagrams?

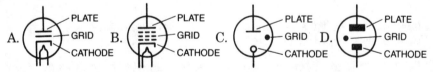

ANSWER A: The triode has three elements on the inside. I know, I know, you do count four elements, but the lower element is the cathode heater, and it is usually considered as part of the cathode element. Notice that the order of the elements are in ascending alphabetical order; that is, "C" for cathode, "G" for grid, and "P" for plate. Remember that "tri" means three, like a triangle is a three-angle figure.

Subelement 2G – Practical Circuits (2 examination questions from 20 questions in 2G)

One question must be from the following 10 questions:

2G-1-1.1 What is the unlabeled block (?) in this diagram?
A. A terminal-node controller
B. An antenna switch
C. A telegraph key
D. A TR switch

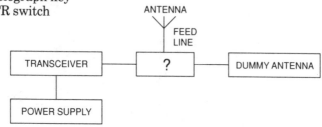

ANSWER B: Look closely at the diagram. You see two antenna systems. One is a dummy antenna, and the other one is a symbol of an antenna. The question mark is a single-pole double-throw antenna switch.

2G-1-1.2 What is the unlabeled block (?) in this diagram?
A. A microphone
B. A receiver
C. A transmitter
D. An SWR meter

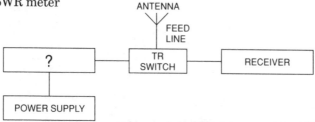

ANSWER C: This is your basic radio. If the receiver is on the right, then you can expect the transmitter to be on the left. The TR switch switches the antenna between transmitter and receiver.

2G-1-1.3 What is the unlabeled block (?) in this diagram?
A. A key click filter
B. An antenna tuner
C. A power supply
D. A receiver

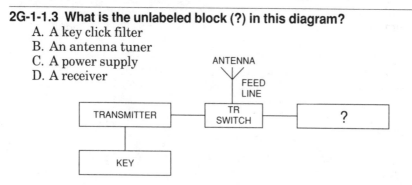

ANSWER D: This time they left out the receiver. The missing block is the receiver. The TR switch has the antenna connected to the receiver when you want to receive signals.

2G-1-1.4 What is the unlabeled block (?) in this diagram?
- A. A transceiver
- B. A TR switch
- C. An antenna tuner
- D. A modem

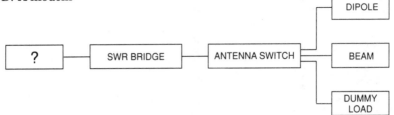

ANSWER A: Here we see the complete set-up, but minus the radio. A radio with a combined transmitter and receiver is called a transceiver. Separate transmitters and receivers are no longer manufactured for ham radio use: both are combined into one integrated unit—the transceiver.

2G-1-1.5 In block Diagram 2G-1, which symbol represents an antenna?
- A. 1
- B. 2
- C. 3
- D. 4

Diagram 2G-1

ANSWER D: It's No. 4, the antenna symbol; no question about it!

2G-1-2.1 What is the unlabeled block (?) in this diagram?
- A. A pi network
- B. An antenna switch
- C. A key click filter
- D. A mixer

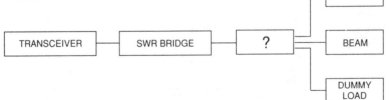

ANSWER B: Here we see three antennas connected to one radio and one SWR bridge. The block in question is a 3-position switch.

2G-1-2.2 What is the unlabeled block (?) in this diagram?
A. A TR switch
B. A variable frequency oscillator
C. A linear amplifier
D. A microphone

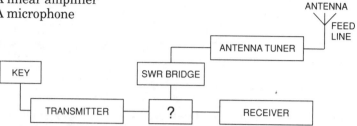

ANSWER A: The TR switch toggles between transmitter and receiver as you key your system.

2G-1-2.3 What is the unlabeled block (?) in this diagram?
A. An antenna switch
B. An impedance-matching network
C. A key click filter
D. A terminal-node controller

ANSWER B: Anything in between the antenna and the standing wave ratio bridge is probably some sort of antenna tuner. Some of today's worldwide radios have built-in automatic antenna tuners. A technical name for an automatic antenna tuner is an "impedance-matching network."

2G-1-2.4 In block diagram 2G-1, if component 1 is a transceiver and component 2 is an SWR meter, what is component 3?
A. A power supply
B. A receiver
C. A microphone
D. An impedance matching device

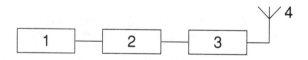

Diagram 2G-1

ANSWER D: The antenna tuner always follows the SWR meter, and precedes the actual antenna system.

2G-1-2.5 In block diagram 2G-1, if component 2 is an SWR meter and component 3 is an impedance matching device, what is component 1?
    A. A power supply
    B. An antenna
    C. An antenna switch
    D. A transceiver
ANSWER D: It takes a transceiver to drive the SWR bridge, and then the matching device, and then the antenna system.

One question must be from the following 10 questions:

2G-2.1 In an amateur station designed for Morse radiotelegraph operation, what station accessory will you need to go with your transmitter?
    A. A terminal-node controller
    B. A telegraph key
    C. An SWR meter
    D. An antenna switch
ANSWER B: Read this question carefully. Since they're asking about radiotelegraph, you would need a telegraph key. Don't confuse "radiotelegraph" and "radiotelephone." For radiotelephone operation, you need a microphone.

2G-2.2 What is the unlabeled block (?) in this diagram of a Morse telegraphy station?
    A. A sidetone oscillator
    B. A microphone
    C. A telegraph key
    D. A DTMF keypad

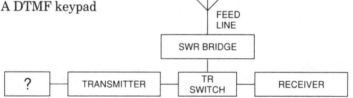

ANSWER C: There it is again, the telegraph key. We can figure this out because they asked for the missing device in the "telegraph" station.

2G-2.3 What station accessory do many amateurs use to help form good Morse code characters?
    A. A sidetone oscillator
    B. A key-click filter
    C. An electronic keyer
    D. A DTMF keypad
ANSWER C: If you enjoy the Morse code, you may wish to invest in an electronic keyer with side paddles. This lets you form perfect dits and dahs.

2G-3.1 In an amateur station designed for radiotelephone operation, what station accessory will you need to go with your transmitter?
    A. A splatter filter
    B. A terminal-voice controller
    C. A receiver audio filter
    D. A microphone

ANSWER D: The key word here is "radiotelephone." Again, read the question carefully so you don't mistakenly read "radiotelegraph." To talk on the telephone, you need a microphone.

2G-3.2 What is the unlabeled block (?) in this diagram of a radiotelephone station?
   A. A splatter filter
   B. A terminal-voice controller
   C. A receiver audio filter
   D. A microphone

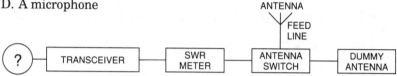

ANSWER D: Here we have a radiotelephone station. For a telephone station, you need a microphone. Makes sense, doesn't it?

2G-4.1 In an amateur station designed for radioteletype operation, what station accessories will you need to go with your transmitter?
   A. A modem and a teleprinter or computer system
   B. A computer, a printer and a RTTY refresh unit
   C. A terminal-node controller
   D. A modem, a monitor and a DTMF keypad

ANSWER A: If you like computers, you are going to love your new Novice computer privileges! Chances are you already own most of the components for going on radioteleprinter, or packet operation. Your present personal computer and printer may need only to be connected to a modem or terminal-node controller (TNC). These easily plug into your ham radio set, and you are on the air!

2G-4.2 What is the unlabeled block (?) in this diagram?
   A. An RS-232 interface
   B. SWR bridge
   C. Modem
   D. Terminal-network controller

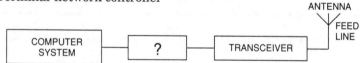

ANSWER C: As you can see, it's not complicated to tie in your ham set for digital communications. Many manufacturers offer transceiver tie-in kits that allow you to interface your set and your computer without ever having to pick up a screwdriver! Be careful here because Answer D says "terminal-network controller" not "terminal-node controller."

2G-5.1 In a packet-radio station, what device connects between the radio transceiver and the computer terminal?
   A. A terminal-node controller
   B. An RS-232 interface

    C. A terminal refresh unit
    D. A tactical network control system

ANSWER A: The heart of your packet station will be a microprocessor-based device called a terminal-node (not network) controller (TNC). It allows you to use your computer on the ham airwaves. Its function is to assemble and disassemble the packets of information. There may be several models available for any type of worldwide or VHF/UHF ham set. If you live on a mountain top, your TNC can also turn your station into a digipeater for the automatic relay of packet transmissions.

2G-5.2 What is the unlabeled block (?) in this diagram of a packet-radio station?
    A. A terminal-node controller
    B. An RS-232 interface
    C. A terminal refresh unit
    D. A tactical network control system

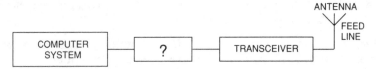

ANSWER A: The terminal-node controller interfaces your computer to your ham set so that you can automatically receive computer messages via Amateur Radio. Most controllers just plug into your computer without any soldering or tools.

2G-5.3 Where does a terminal-node controller connect in an amateur packet-radio station?
    A. Between the antenna and the radio
    B. Between the computer and the monitor
    C. Between the computer or terminal and the radio
    D. Between the keyboard and the computer

ANSWER C: It's important to tell the salesperson the type of ham set and type of computer you have in order to get the right TNC and interconnecting cables.

Subelement 2H – Signals and Emissions (2 examination questions from 23 questions in 2H)

One question must be from the following 12 questions:

2H-1-1.1 What keying method is used to transmit CW?
    A. Frequency-shift keying of a radio-frequency signal
    B. On/off keying of a radio-frequency signal
    C. Audio-frequency-shift keying of an oscillator tone
    D. On/off keying of an audio-frequency signal

ANSWER B: The telegraph key turns on and off the radio frequency signal. CW is A1A, or Morse code. This signal has no modulation—it's just interrupted carrier CW (continuous wave). The carrier is on for the duration of a "dit" or "dah" and is off the rest of the time.

2H-1-1.2 What emission type describes international Morse code telegraphy messages?
    A. RTTY
    B. Image
    C. CW
    D. Phone
ANSWER C: CW stands for continuous wave, another name for Morse Code. In the emission table at Question 2A-17.2, it would be A1A.

2H-1-2.1 What emission type describes narrow-band direct-printing telegraphy emissions?
    A. RTTY
    B. Image
    C. CW
    D. Phone
ANSWER A: Radioteleprinting is a popular Amateur Radio passtime. RTTY stands for radioteletype. In the emission table at Question 2A-17.2, it would be F1B.

2H-1-2.2 What keying method is used to transmit RTTY messages?
    A. Frequency-shift keying of a radio-frequency signal
    B. On/off keying of a radio-frequency signal
    C. Digital pulse-code keying of an unmodulated carrier
    D. On/off keying of an audio-frequency signal
ANSWER A: This is many times abbreviated FSK, and it's radioteletype. Here the frequency of the carrier is shifted to identify the information being transmitted.

2H-1-3.1 What emission type describes frequency-modulated voice transmissions?
    A. FM phone
    B. Image
    C. CW
    D. Single-sideband phone
ANSWER A: When you see the word "phone", it always means voice. In the emission table at Question 2A-17.2, it would be F3E.

2H-1-4.1 What emission designator describes single-sideband suppressed-carrier (SSB) voice transmissions?
    This question has been retracted by the QPC because the FCC established, effective February 14, 1991, the Technician Class as a no-code license class.
    This question is not to be used on any written examination administered after February 14, 1991 until or unless it is revised or replaced by the QPC.

2H-2.1 What does the term *key click* mean?
    A. The mechanical noise caused by a closing a straight key too hard
    B. The clicking noise from an excessively square CW keyed waveform
    C. The sound produced in a receiver from a CW signal faster than 20 WPM
    D. The sound of a CW signal being copied on an AM receiver
ANSWER B: If you pass a code test, we suggest you subscribe to the America Radio Relay League's *QST* magazine which also makes you a member of the

American Radio Relay League. In the magazine, you will see product comparisons. They often show oscilloscope pictures of the keying waveform in the CW mode. Smooth CW has a nice curve to the starting of the signal that will minimize key clicks. Key clicks are the tapping sounds of a CW signal that you can hear all over the band without actually hearing the tone.

2H-2.2 How can key clicks be eliminated?
A. By reducing your keying speed to less than 20 WPM
B. By increasing power to the maximum allowable level
C. By using a power supply with better regulation
D. By using a key-click filter

ANSWER D: Unless you build your own gear, you shouldn't have a problem with key clicks. Modern transceivers for worldwide use all feature built-in key click filters. All you need to do is to buy a key, plug it in, and have fun with CW.

2H-3.1 What does the term *chirp* mean?
A. A distortion in the receiver audio circuits
B. A high-pitched audio tone transmitted with a CW signal
C. A slight shift in oscillator frequency each time a CW transmitter is keyed
D. A slow change in transmitter frequency as the circuit warms up

ANSWER C: Chirps are changes in frequency at the beginning of a keying pulse and are usually caused by poor power supply voltage regulation. Many worldwide ham sets operate only from 12 volts. This keeps their size down. For home use, you will need at least a 20-amp power supply to run your new worldwide 10-meter setup. If you try to use an old CB-type power supply, chances are it will pull the voltage down so low that your transmitter's keyed tone will wobble, or chirp, in frequency.

2H-3.2 What can be done to the power supply of a CW transmitter to avoid chirp?
A. Resonate the power supply filters.
B. Regulate the power supply output voltages.
C. Use a buffer amplifier between the transmitter output and the feed line.
D. Hold the power supply current to a fixed value.

ANSWER B: A signal with chirp is an unstable signal. Low voltage is many times the culprit. A better regulation of the power supply output voltage will cure the problem. If you are running on batteries, charge the batteries!

2H-4.1 What is a common cause of superimposed hum?
A. Using a nonresonant random-wire antenna
B. Sympathetic vibrations from a nearby transmitter
C. Improper neutralization of the transmitter output stage
D. A defective filter capacitor in the power supply

ANSWER D: If you find a 20-year-old power supply your dad used back in the good old days for his ham radio setup, better check its dc output power. Very old power supplies feature big electrolytic filter capacitors that dry out and lose their capacity with age. The result will be severe hum on transmit and receive. Power supply technology has changed so rapidly that switching power supplies can easily deliver 12 volts at 20 amps in small and relatively lightweight packages.

They have done away with the very heavy transformers needed in the older conventional power supplies. The switching power supply also doesn't have those big filter capacitors that dry out.

2H-4.2 What type of problem can a bad power-supply filter capacitor cause in a transmitter or receiver?
   A. Sympathetic vibrations in nearby receivers
   B. A superimposed hum or buzzing sound
   C. Extreme changes in antenna resonance
   D. Imbalance in the mixers
ANSWER B: The big filter capacitors in older power supplies sometimes begin to dry out. If everyone says your signal has hum on it, plan to change those capacitors or get a modern power supply. Also, if you are accused of hum, make sure you are not transmitting while plugged into the wall and charging your batteries.

One question must be from the following 11 questions:

2H-5.1 What is the 4th harmonic of a 7160-kHz signal?
   A. 28,640 kHz
   B. 35,800 kHz
   C. 28,160 kHz
   D. 1790 kHz
ANSWER A: Harmonics are multiples of your transmitted signal. Simply multiply the harmonic number times the fundamental frequency. In this case, multiply 4 times 7160 kHz and the result is Answer A. Calculator keystrokes are: CLEAR 4 $\times$ 7160 = and the answer is 28640.

2H-5.2 You receive an FCC *Notice of Violation* stating that your station was heard on 21,375 kHz. At the time listed on the notice, you were operating on 7125 kHz. What is a possible cause of this violation?
   A. Your transmitter has a defective power-supply filter capacitor
   B. Your CW keying speed was excessively fast
   C. Your transmitter was radiating excess harmonic signals
   D. Your transmitter has a defective power-supply filter choke
ANSWER C: Because these two frequencies are harmonically related, chances are your ham set is an older model and is generating harmonics. You need to work on the radio, or get it fixed professionally. Newer ham sets rarely have this problem.

2H-6.1 What may happen to body tissues that are exposed to large amounts of UHF or microwave RF energy?
   A. The tissue may be damaged because of the heat produced.
   B. The tissue may suddenly be frozen.
   C. The tissue may be immediately destroyed because of the Maxwell Effect.
   D. The tissue may become less resistant to cosmic radiation.
ANSWER A: Welcome back to our microwave oven and RF emissions theory. As much as possible, keep your head, especially your eyes, away from the radiating antennas of VHF/UHF transceivers.

2H-6.2 What precaution should you take before working near a high-gain UHF or microwave antenna (such as a parabolic, or dish antenna)?
A. Be certain the antenna is FCC type accepted.
B. Be certain the antenna and transmitter are properly grounded.
C. Be certain the transmitter cannot be operated.
D. Be certain the antenna safety interlocks are in place.
ANSWER C: One way of achieving antenna gain in one direction is to focus the 1270-MHz energy with a microwave dish antenna. Never stand in front of a dish antenna when the ham set is in operation.

2H-6.3 You are installing a VHF or UHF mobile radio in your vehicle. What is the best location to mount the antenna on the vehicle to minimize any danger from RF exposure to the driver or passengers?
A. In the middle of the roof
B. Along the top of the windshield
C. On either front fender
D. On the trunk lid
ANSWER A: Mounting it in the middle of the roof provides maximum isolation from RF radiating from the antenna. However, this is usually not practical. Trunk-lid mounts and rear-glass mounts are the next best choice.

2H-7.1 You discover that your tube-type transmitter power amplifier is radiating spurious emissions. What is the most likely cause of this problem?
A. Excessively fast keying speed
B. Undermodulation
C. Improper neutralization
D. Tank-circuit current dip at resonance
ANSWER C: When we neutralize an older tube-type transmitter, we neutralize out signals on all frequencies except for the fundamental frequency of the desired signal. These non-harmonic-related frequencies generated by poorly neutralized radios are called spurious emissions. They can fall on hundreds of frequencies throughout the spectrum. Neutralization is the way to get rid of them.

2H-7.2 Your transmitter radiates signals outside the amateur band where you are transmitting. What term describes this radiation?
A. Off-frequency emissions
B. Transmitter chirp
C. Incidental radiation
D. Spurious emissions
ANSWER D: Here again the question says "signals," so they are surely spurious emissions.

2H-7.3 What problem can occur if you operate your transmitter without the cover and other shielding in place?
A. Your transmitter can radiate spurious emissions.
B. Your transmitter may radiate a "chirpy" signal.
C. The final amplifier efficiency of your transmitter may decrease.
D. You may cause splatter interference to other stations operating on nearby frequencies.

ANSWER A: If you take your transmitter apart, make sure you completely reassemble it. If you leave out any metal covers, chances are it could radiate spurious emissions.

2H-7.4 What type of interference will you cause if you operate your SSB transmitter with the microphone gain adjusted too high?
   A. You may cause digital interference to computer equipment in your neighborhood.
   B. You may cause splatter interference to other stations operating on nearby frequencies.
   C. You may cause atmospheric interference in the air around your antenna.
   D. You may cause processor interference to the microprocessor in your rig.

ANSWER B: Worldwide transceivers all have microphone gain controls. Many base station microphones also offer audio compression and additional mike gain. Turning up the mike gain too high will cause distortion, splatter, and interference to stations on nearby frequencies. It's also bad Amateur Radio practice to operate with a "hot" mike setting. Read your instruction manual for proper mike gain setting procedures while watching the ALC level on your transceiver's multimeter.

2H-7.5 What may happen if you adjust the microphone gain or deviation control on your FM transmitter too high?
   A. You may cause digital interference to computer equipment in your neighborhood.
   B. You may cause interference to other stations operating on nearby frequencies.
   C. You may cause atmospheric interference in the air around your antenna.
   D. You may cause processor interference to the microprocessor in your rig.

ANSWER B: On your FM equipment, there is no microphone gain control. It has been preset at the factory, so you probably won't ever need to worry about microphone gain when operating with a stock microphone. However, if you decide to change to a different type of microphone, you may wish to let a ham radio expert check the audio level.

2H-7.6 What type of interference can excessive amounts of speech processing in your SSB transmitter cause?
   A. You may cause digital interference to computer equipment in your neighborhood.
   B. You may cause splatter interference to other stations operating on nearby frequencies.
   C. You may cause atmospheric interference in the air around your antenna.
   D. You may cause processor interference to the microprocessor in your rig.

ANSWER B: Modern worldwide transceivers, such as the set you will use on 10 meters, contain powerful speech processing circuits that intensify your transmitted voice signal. Too high a setting of the speech processor will lead to splatter and terrible sounding transmissions. In fact, avoid using the speech processor unless absolutely necessary. Using the speech processor widens your transmitted signal, possibly causing interference to stations operating on frequencies close to yours.

Subelement 2I – Antennas and Feed Lines (3 examination questions from 39 questions in 2I)

One question must be from the following 12 questions:

2I-1.1 What is the approximate length (in feet) of a half-wavelength dipole antenna for 3725 kHz?
   A. 126 ft
   B. 81 ft
   C. 63 ft
   D. 40 ft

ANSWER A: A simple antenna to build for worldwide operation is called the half-wave dipole. This simple antenna contains everyday wire whose length is one-half of the wavelength of the frequency being transmitted. The signal from your transmitter is applied in the exact middle of the dipole. It is best to use coaxial cable for this feed-in connection.

The length of a half-wave dipole is easily calculated using the equation:

$$\lambda = \frac{468}{f}$$

where λ is the wavelength in feet and f is the frequency in MHz.

To use the equation, you must convert kilohertz to megahertz. Move the decimal point three places to the left to convert 3725 kHz to 3.725 MHz. Now divide 3.725 into 468. The calculator keystrokes are: CLEAR 468 ÷ 3.725 = and the answer is 125.6 feet. Round it to 126 feet to match the answer choice. Try it, it's easy! Remember 468 to calculate the length in feet of a half-wave dipole.

2I-1.2 What is the approximate length (in feet) of a half-wavelength dipole antenna for 7125 kHz?
   A. 84 ft
   B. 42 ft
   C. 33 ft
   D. 66 ft

ANSWER D: See Question 2I-1.1. The higher you go in frequency, the shorter the dipole. Remember to change kilohertz to megahertz (7.125 MHz) and divide the result into 468. Dipoles operate best when they are at least 30 feet above the earth. They work great on a roof. Calculator keystrokes are: CLEAR 468 ÷ 7.125 = and the answer is 65.7. Round it to 66 to match the answer choice.

2I-1.3 What is the approximate length (in feet) of a half-wavelength dipole antenna for 21,125 kHz?
   A. 44 ft
   B. 28 ft
   C. 22 ft
   D. 14 ft

ANSWER C: See Question 2I-1.1. Again, the higher you go in frequency, the shorter the dipole length. Convert kilohertz to megahertz by moving the decimal point three places to the left (21.125 MHz), and divide into 468. Calculator keystrokes are: CLEAR 468 ÷ 21.125 = and the answer is 22.2. Round it to 22 to match the answer choice.

2I-1.4 What is the approximate length (in feet) of a half-wavelength dipole antenna for 28,150 kHz?
  A. 22 ft
  B. 11 ft
  C. 17 ft
  D. 34 ft
ANSWER C: Here we are at the 10-meter band, 28 MHz. This is where your new privileges allow voice, so this dipole will be quite popular. There is good news; after you calculate its length, (see Question 2I-1.1.), you will see that it's only 17 feet long. With this antenna you can go almost anywhere! Calculator keystrokes are: CLEAR 468 ÷ 28.15 = and the answer is 16.6. Round it to 17 to match the answer choice.

2I-1.5 How is the approximate length (in feet) of a half-wavelength dipole antenna calculated?
  A. By substituting the desired operating frequency for f in the formula: 150/ f (in MHz)
  B. By substituting the desired operating frequency for f in the formula: 234/ f (in MHz)
  C. By substituting the desired operating frequency for f in the formula: 300/ f (in MHz)
  D. By substituting the desired operating frequency for f in the formula: 468/ f (in MHz)
ANSWER D: Plug this into your memory—you will use it often in your ham radio career.

2I-2.1 What is the approximate length (in feet) of a quarter-wavelength vertical antenna for 3725 kHz?
  A. 20 ft
  B. 32 ft
  C. 40 ft
  D. 63 ft
ANSWER D: For mobile operation, we use a variation of the dipole that is vertically polarized. We use the vehicle frame as a portion of the dipole antenna. Half of the dipole is your vehicle, and the other half is the whip. This whip is exactly one-quarter wavelength of the frequency used for operation. Now don't get confused—to calculate the length of a quarter wave mobile whip, simply take the length in feet you figured out for a half-wave dipole and *cut it in half*. That's all there is to it. Don't forget to convert frequency to MHz. Calculator keystrokes are: CLEAR 468 ÷ 3.725 ÷ 2 = and the answer is 62.8. Round it to 63.

2I-2.2 What is the approximate length (in feet) of a quarter-wavelength vertical antenna for 7125 kHz?
  A. 11 ft
  B. 16 ft

C. 21 ft

D. 33 ft

ANSWER D: 7125 kHz is 7.125 MHz. Divide 7.125 MHz into 468, and then that result by two. A quarter-wave antenna is just half the length of a half-wave dipole. If you made the calculation correctly, your answer comes out to 33 feet. But you've never seen an antenna that long on a car, have you? This is why worldwide ham antennas for cars use loading coils to shorten their physical length, but have the correct electrical length. If you have antenna restrictions in your neighborhood, these "loaded" worldwide antennas also work quite nicely in an attic. Calculator keystrokes are: CLEAR 468 ÷ 7.125 ÷ 2 = and the answer is 32.8. Round it to 33.

2I-2.3 What is the approximate length (in feet) of a quarter-wavelength vertical antenna for 21,125 kHz?

A. 7 ft

B. 11 ft

C. 14 ft

D. 22 ft

ANSWER B: 21,125 kHz is 21.125 MHz. Divide 21.125 into 468, and then halve the result to obtain your answer. Calculator keystrokes are: CLEAR 468 ÷ 21.125 ÷ 2 = and the answer is 11.1. Round it to 11. This quarter-wave whip doesn't require much loading at all—it's only 11 feet long.

2I-2.4 What is the approximate length (in feet) of a quarter-wavelength vertical antenna for 28,150 kHz?

A. 5 ft

B. 8 ft

C. 11 ft

D. 17 ft

ANSWER B: 28.150 divided into 468, and then take half the answer to end up with a quarter-wave vertical whip. You have seen many of these mobile quarter-wave whips running around in town, and this is what CB radio operators use. This same stainless steel whip, slightly shorter for the ham band, works great on 10 meters for 10 meter mobile worldwide contacts. You can also buy shortened versions of the 10 meter worldwide antenna that are base-loaded. Calculator keystrokes are: CLEAR 468 ÷ 28.15 ÷ 2 = and the answer is 8.3. Round it to 8.

2I-2.5 When a vertical antenna is lengthened, what happens to its resonant frequency?

A. It decreases

B. It increases

C. It stays the same

D. It doubles

ANSWER A: Remember the rule we discussed previously: "lower-longer." If an antenna is cut a little bit too long, it works better lower in frequency.

2I-3.1 Why do many amateurs use a 5/8-wavelength vertical antenna rather than a 1/4-wavelength vertical antenna for their VHF or UHF mobile stations?

2I–ANTENNAS AND FEED LINES

A. A 5/8-wavelength antenna can handle more power than a 1/4-wavelength antenna.

B. A 5/8-wavelength antenna has more gain than a 1/4-wavelength antenna.

C. A 5/8-wavelength antenna exhibits less corona loss than a 1/4-wavelength antenna.

D. A 5/8-wavelength antenna looks more like a CB antenna, so it does not attract as much attention as a 1/4-wavelength antenna.

ANSWER B: There are about ten manufacturers of mobile VHF and UHF ham antennas. Some are trunk-lip mounted, some go through the car chassis (ugh), and some will even hang onto the window glass. A 5/8-wave antenna is much taller than the quarterwave antenna, but it has more gain.

2I-3.2 What type of radiation pattern is produced by a 5/8-wavelength vertical antenna?

A. A pattern with most of the transmitted signal concentrated in two opposite directions

B. A pattern with the transmitted signal going equally in all compass directions, with most of the radiation going high above the horizon

C. A pattern with the transmitted signal going equally in all compass directions, with most of the radiation going close to the horizon

D. A pattern with more of the transmitted signal concentrated in one direction than in other directions

ANSWER C: A dipole antenna radiates the signal bidirectionally—broadside to the wire. A vertical antenna radiates the signal vertically equally in all directions.

One question must be from the following 10 questions:

2I-4-1.1 What type of antenna produces a radiation pattern with more of the transmitted signal concentrated in a particular direction than in other directions?

A. A dipole antenna

B. A vertical antenna

C. An isotropic antenna

D. A beam antenna

ANSWER D: Another name for a beam antenna is a Yagi antenna. You can easily convert your CB 11-meter Yagi antenna into a powerful 10-meter beam.

2I-4-1.2 What type of radiation pattern is produced by a Yagi antenna?

A. A pattern with the transmitted signal spread out equally in all compass directions

B. A pattern with more of the transmitted signal concentrated in one direction than in other directions

C. A pattern with most of the transmitted signal concentrated in two opposite directions

D. A pattern with most of the transmitted signal concentrated at high radiation angles

ANSWER B: Dr. H. Yagi, a Japanese physicist, translated into English the directional beam antenna design named after him. Actually, a professor named Uda invented the antenna with the placement of a slightly shorter dipole in front of, and slightly longer dipole in back of, a regular dipole to cause the signal to concentrate in one direction. The longer element reflected some of the energy and the shorter element directed the energy in one specific direction. So the Yagi antenna is really a series of slightly longer and shorter dipoles that influence a directly fed dipole to transmit better in a forward direction. (See Figure 2I-4)

2I-4-1.3 Approximately how long (in wavelengths) is the driven element of a Yagi antenna?
A. 1/4 wavelength
B. 1/3 wavelength
C. 1/2 wavelength
D. 1 wavelength

ANSWER C: Since the Yagi antenna is a series of dipoles affixed to a boom in the same plane, the individual elements are about one-half wavelength long, just like the driven element. The reflector is a little longer, the director a little shorter.

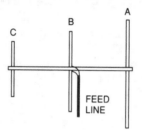

Figure 2I-4

2I-4-2.1 On the Yagi antenna shown in Figure 2I-4, what is the name of section B?
A. Director
B. Reflector
C. Boom
D. Driven element

ANSWER D: The coaxial cable always connects to the driven element. It is driven by the direct energy from the transmitter.

2I-4-2.2 On the Yagi antenna shown in Figure 2I-4, what is the name of section C?
A. Director
B. Reflector
C. Boom
D. Driven element

ANSWER A: The director is usually the shortest element and is placed in front of the driven element. Directors help concentrate the signal into a tight radiation pattern.

2I-4-2.3 On the Yagi antenna shown in Figure 2I-4, what is the name of section A?
A. Director
B. Reflector
C. Boom
D. Driven element

ANSWER B: The reflector is on the back of the beam. A reflector is always longer than the driven element. Sometimes there are two or more reflectors.

2I-4-2.4 What are the names of the elements in a 3-element Yagi antenna?
A. Reflector, driven element and director
B. Boom, mast and reflector
C. Reflector, base and radiator
D. Driven element, trap and feed line
ANSWER A: Be careful with this one—they're asking for the elements of a Yagi, and the boom is not included in the count. By the way, the longer the boom, the more elements and the more powerful the Yagi.

2I-5.1 How should the antenna on a handheld transceiver be positioned while you are transmitting?
A. Away from your head and away from others standing nearby
B. Pointed in the general direction of the repeater or other station you are transmitting to
C. Pointed in a general direction 90 degrees away from the repeater or other station you are transmitting to
D. With the top of the antenna angled down slightly to take the most advantage of ground reflections
ANSWER A: Even little handheld radios could pose a radiation hazard if the antenna is too close to your eyes. The key word here is "handheld."

2I-5.2 Why should you always locate your antennas so that no one can come in contact with them while you are transmitting?
A. Such contact can detune the antenna, causing television interference.
B. To prevent RF burns and excessive exposure to RF energy
C. The antenna is more likely to radiate harmonics when it is touched.
D. Such contact may reflect the transmitted signal back to the transmitter, damaging the final amplifier.
ANSWER B: It makes good sense to always keep antennas away from someone that might touch them.

2I-5.3 You are going to purchase a new antenna for your VHF or UHF handheld radio. Which type of antenna is the best choice to produce a radiation pattern that will be least hazardous to your face and eyes?
A. A 1/8-wavelength whip
B. A 7/8-wavelength whip
C. A 1/2-wavelength whip
D. A short, helically wound, flexible antenna
ANSWER C: You can buy, as an inexpensive accessory, telescopic metal whips with a tiny base loading coil for your handheld. They will add dramatic range to your communications. The halfwave whip is most common, and puts the radiation well above your head.

One question must be from the following 17 questions:

2I-6.1 What is a coaxial cable?
A. Two parallel conductors encased along the edges of a flat plastic ribbon
B. Two parallel conductors held at a fixed distance from each other by insulating rods

C. Two conductors twisted around each other in a double spiral
D. A center conductor encased in insulating material which is covered by a conducting sleeve or shield

ANSWER D: Coaxial cable is similar to water pipes in that it doesn't leak and carries the good stuff inside while keeping out everything else. To maintain high water pressure at the output end requires large diameter pipes. Same thing with coaxial cable—if you plan to run a lot of power, or if you have an extremely long cable run, use a large diameter coaxial cable.

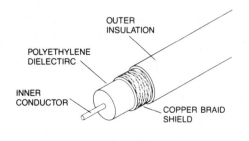

Coaxial Cable

2I-6.2 What kind of antenna feed line is constructed of a center conductor encased in insulation which is then covered by an outer conducting shield and weatherproof jacket?
A. Twin lead
B. Coaxial cable
C. Open-wire feed line
D. Wave guide

ANSWER B: See figure in question 2I-6.1. Buy your coaxial cable from a reputable electronics dealer. While coaxial cable may look the same from the outside, its percentage of braid and gauge of center conductor may vary dramatically. If your coaxial run is not exposed to the elements, plastic jacket cable (RG-8U) is fine; however, if you plan to run a lot of coaxial cable on the roof where it will be bombarded by ultraviolet radiation from the sun, it's best to use a PVC non-contaminating jacket (RG-213).

2I-6.3 What are some advantages of using coaxial cable as an antenna feed line?
A. It is easy to make at home, and it has a characteristic impedance in the range of most common amateur antennas
B. It is weatherproof, and it has a characteristic impedance in the range of most common amateur antennas
C. It can be operated at a higher SWR than twin lead, and it is weatherproof
D. It is unaffected by nearby metallic objects, and has a characteristic impedance that is higher than twin lead

ANSWER B: You can even bury non-contaminating quality coaxial cable. Your author's station uses coaxial cable exclusively, and external runs are underground in PVC tubes so the gophers can't get to the cable.

2I-6.4 What commonly-available antenna feed line can be buried directly in the ground for some distance without adverse effects?
    A. Twin lead
    B. Coaxial cable
    C. Parallel conductor
    D. Twisted pair
ANSWER B: It's very important to completely weatherproof the antenna connection that is out in the open. Coaxial cable connectors may leak moisture, so seal them well.

2I-6.5 When an antenna feed line must be located near grounded metal objects, which commonly-available feed line should be used?
    A. Twisted pair
    B. Twin lead
    C. Coaxial cable
    D. Ladder-line
ANSWER C: Since the shield is at ground potential, running coaxial cable close to grounded objects will not affect the signal. If you need to hide your coaxial cable on a run from the ground floor to the roof, consider running it inside a rain gutter's down-spout.

2I-7.1 What is *parallel-conductor* feed line?
    A. Two conductors twisted around each other in a double spiral
    B. Two parallel conductors held a uniform distance apart by insulating material
    C. A conductor encased in insulating material which is then covered by a conducting shield and a weatherproof jacket
    D. A metallic pipe whose diameter is equal to or slightly greater than the wavelength of the signal being carried
ANSWER B: A less-used feedline is "ladder line." It's similar to TV twin-lead. Ladder line can be used with certain types of wire antenna systems. Ladder line cannot be connected directly to any ham set; it must have an additional antenna tuning device.

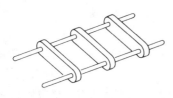

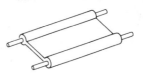

Parallel-Conductor   Twin Lead

2I-7.2 How can *TV-type twin lead* be used as a feed line?
    A. By carefully running the feed line parallel to a metal post to ensure self resonance
    B. TV-type twin lead cannot be used in an amateur station
    C. By installing an impedance-matching network between the transmitter and feed line
    D. By using a high-power amplifier and installing a power attenuator between the transmitter and feed line

ANSWER C: The antenna tuning device is sometimes called a "transmatch." It must be adjusted each time you change bands or make big frequency changes.

2I-7.3 What are some *advantages* of using parallel-conductor feed line?

A. It has a lower characteristic impedance than coaxial cable, and will operate at a higher SWR than coaxial cable.
B. It will operate at a higher SWR than coaxial cable, and it is unaffected by nearby metal objects.
C. It has a lower characteristic impedance than coaxial cable, and has less loss than coaxial cable.
D. It will operate at higher SWR than coaxial cable and it has less loss than coaxial cable.

ANSWER D: There was a ham that lived in a valley who put a wire antenna several thousand feet away on the top of a hill. He used parallel conductor feedline to connect his set with the hilltop antenna because there is less loss in this type of open-wire feed system than in coaxial cable.

2I-7.4 What are some *disadvantages* of using parallel-conductor feed line?

A. It is affected by nearby metallic objects, and it has a characteristic impedance that is too high for direct connection to most amateur transmitters.
B. It is more difficult to make at home than coaxial cable and it cannot be operated at a high SWR.
C. It is affected by nearby metallic objects, and it cannot handle the power output of a typical amateur transmitter.
D. It has a characteristic impedance that is too high for direct connection to most amateur transmitters, and it will operate at a high SWR.

ANSWER A: Parallel-conductor feed line cannot be run beside any metal object. Also, you'll need an antenna matcher to transform its 300- to 600-ohm impedance to your transceiver's 52-ohm impedance.

2I-7.5 What kind of antenna feed line is constructed of two conductors maintained a uniform distance apart by insulated spreaders?

A. Coaxial cable
B. Ladder-line open conductor line
C. Twin lead in a plastic ribbon
D. Twisted pair

ANSWER B: It is sometimes called a *parallel open-wire feedline*. While not very popular with today's amateurs, it can be built by using solid copper wire and plastic spacers.

2I-8.1 A certain antenna has a feed-point impedance of 35 ohms. You want to use a coaxial cable with 50 ohms impedance to feed this antenna. What type of device will you need to connect between the antenna and the feed line?

A. A balun
B. An SWR bridge
C. An impedance matching device
D. A low-pass filter

ANSWER C: Many homemade vertical antennas will have an impedance mismatch. We use a small impedance matching device to match the antenna precisely to the transceiver.

2I-8.2 A certain antenna system has an impedance of 1000 ohms on one band. What must you use to connect this antenna system to the 50-ohm output on your transmitter?
A. A balun
B. An SWR bridge
C. An impedance matching device
D. A low-pass filter

ANSWER C: This big mismatch of impedance is easily taken care of with an antenna tuner, or an impedance matching device.

2I-9.1 The word balun is a contraction for what phrase?
A. Balanced-antenna-lobe use network
B. Broadband-amplifier linearly unregulated
C. Balanced unmodulator
D. Balanced to unbalanced

ANSWER D: We use a balun to match balanced twin-lead to unbalanced 50 ohm coaxial cable.

2I-9.2 Where would you install a balun if you wanted to feed your dipole antenna with 450-ohm parallel-conductor feed line?
A. At the transmitter end of the feed line
B. At the antenna feed point
C. In only one conductor of the feed line
D. From one conductor of the feed line to ground

ANSWER A: If you do feed your balanced dipole with parallel conductor balanced feedline, the balun, in this case, would go right at the transmitter.

2I-9.3 Where might you install a balun if you wanted to feed your dipole antenna with 50-ohm coaxial cable?
A. You might install a balun at the antenna feed point.
B. You might install a balun at the transmitter output.
C. You might install a balun 1/2 wavelength from the transmitter.
D. You might install baluns in the middle of each side of the dipole.

ANSWER A: This is a much more common scenario—coaxial cable feeding a balanced dipole. In this case, the balun is installed on the antenna at the antenna feedpoint.

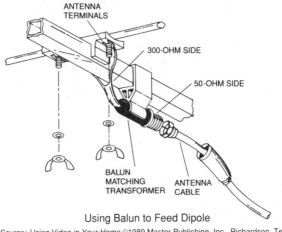

Using Balun to Feed Dipole

Source: *Using Video in Your Home* ©1989 Master Publishing, Inc., Richardson, Texas

2I-10-1.1 A four-element Yagi antenna is mounted with its elements parallel to the ground. A signal produced by this antenna will have what type of polarization?
A. Broadside polarization
B. Circular polarization
C. Horizontal polarization
D. Vertical polarization
ANSWER C: Antenna polarization is just like it looks—if the elements are horizontal to the ground, polarization will be horizontal. Elements parallel to the ground are considered horizontal.

2I-11-1.1 A four-element Yagi antenna is mounted with its elements perpendicular to the ground. A signal produced by this antenna will have what type of polarization?
A. Broadside polarization
B. Circular polarization
C. Horizontal polarization
D. Vertical polarization
ANSWER D: If the elements are perpendicular to the ground, they are vertical. They radiate signals with vertical polarization.

CHECK YOURSELF
At this point, go back and check your work. Put a check mark beside each question and answer you have memorized correctly. Continue to review the questions and answers with which you may be having a little problem. Let a friend read you the correct answer, and see if you can recite the question! Continue to review the questions and answers until you have mastered them so you don't miss any more than 1 in 10. Remember, you will have a 30-question written examination on Element 2, and you must get 22 correct answers. Have someone give you a trial examination to see how you do.

ELEMENT 3A QUESTION POOL
When you are confident that you have mastered Element 2, continue the same study pattern with Element 3A. Here are the 326 Element 3A questions, the correct answers, and an explanation for each correct answer. Continue to study hard, and good luck!

Subelement 3AA – Commission's Rules (5 examination questions from 64 questions in 3AA)

3AA-1.1 What is the *control point* of an amateur station?
A. The location at which the control operator function is performed
B. The operating position of any amateur station operating as a repeater user station
C. The physical location of any Amateur Radio transmitter, even if it is operated by radio link from some other location
D. The variable frequency oscillator (VFO) of the transmitter

ANSWER A: This is where you have complete capabilities to turn the equipment on, or shut it off, in case of a malfunction. Every ham radio station is required to have a control point.

3AA-1.2 What is the term for the location at which the control operator function is performed?
A. The operating desk
B. The control point
C. The station location
D. The manual control location

ANSWER B: Some repeaters that may be used for phone patches are controlled by radio links. This means that the actual control point is with the control operator and his tiny handheld transceiver. When you upgrade to Technician, you too could be a control operator with the control point worn on your belt!

3AA-2.1 What are the HF privileges authorized to a Technician Class control operator in ITU region 2?

This question has been retracted by the QPC because the FCC established, effective February 14, 1991, the Technician Class as a no-code license class.

This question is not to be used on any written examination administered after February 14, 1991 until or unless it is revised or replaced by the QPC.

3AA-2.2 Which operator licenses authorize privileges on 52.525 MHz?
A. Extra, Advanced only
B. Extra, Advanced, General only
C. Extra, Advanced, General, Technician only
D. Extra, Advanced, General, Technician, Novice

ANSWER C: 52.525 Mhz is the middle of the 6-meter wavelength band. Everyone but the Novice has privileges here.

3AA-2.3 Which operator licenses authorize privileges on 146.52 MHz?
A. Extra, Advanced, General, Technician, Novice
B. Extra, Advanced, General, Technician only
C. Extra, Advanced, General only
D. Extra, Advanced only

ANSWER B: 146.52 Mhz is in the middle of the 2-meter wavelength band. Again, everybody but the Novice has privileges here.

3AA-2.4 Which operator licenses authorize privileges on 223.50 MHz?
A. Extra, Advanced, General, Technician, Novice
B. Extra, Advanced, General, Technician only
C. Extra, Advanced, General only
D. Extra, Advanced only

ANSWER A: This is in the middle of the 220 MHz Novice band, and since Novices have voice privileges up here, everyone is allowed on this frequency.

3AA-2.5 Which operator licenses authorize privileges on 446.0 MHz?
A. Extra, Advanced, General, Technician, Novice
B. Extra, Advanced, General, Technician only
C. Extra, Advanced, General only
D. Extra, Advanced only

ANSWER B: The 440-MHz band is off limits to Novices—but to everyone else, 446.0 MHz is a fun spot for UHF operating.

3AA-3.1 How often do amateur service licenses generally need to be renewed?
A. Every 10 years
B. Every 5 years
C. Every 2 years
D. They are lifetime licenses

ANSWER A: Now longer than 1 year, now longer than 2 years, now longer than 5 years, now 10 years! This is the longest term in history that we have ever had for our amateur services licenses.

3AA-3.2 The FCC currently issues amateur licenses carrying 10-year terms. What is the "grace period" during which the FCC will renew an expired 10-year license?
A. 2 years
B. 5 years
C. 10 years
D. There is no grace period.

ANSWER A: You are not allowed to operate during a grace period; however, you can keep your privileges for 2 years. After that, they are lost for good. So is your call sign. Don't forget to renew!

3AA-3.3 What action would you take to modify your operator/primary station license?
A. Properly fill out FCC Form 610 and send it to the FCC in Gettysburg, PA.
B. Properly fill out FCC Form 610 and send it to the nearest FCC field office.

C. Write the FCC at their nearest field office.

D. There is no need to modify an amateur license between renewals.

ANSWER A: Don't write a letter! All modifications and renewals are done on FCC Form 610. There is a Form 610 bound into the back of this book.

3AA-4.1 On what frequencies within the 6-meter band may FM phone emissions be transmitted?

A. 50.0-54.0 MHz only

B. 50.1-54.0 MHz only

C. 51.0-54.0 MHz only

D. 52.0-54.0 MHz only

ANSWER B: In emission classification, F3E stands for FM voice modulation, and it's allowed throughout the 6-meter wavelength band except below 50.1 to 50.0 MHz.

3AA-4.2 On what frequencies within the 2-meter band may FM image emissions be transmitted?

A. 144.1-148.0 MHz only

B. 146.0-148.0 MHz only

C. 144.0-148.0 MHz only

D. 146.0-147.0 MHz only

ANSWER A: Frequency modulated TV (F3F) is allowed on the 2-meter wavelength band from 144.1 MHz to 148.0 MHz and you have FM voice privileges (F3E) throughout the band except below 144.1 MHz to 144.0 MHz.

3AA-4.3 What emission type may always be used for station identification, regardless of the transmitting frequency?

A. CW

B. RTTY

C. MCW

D. Phone

ANSWER A: CW stands for continuous wave. We use an interrupted continuous wave to transmit the dots and dashes of telegraphy. Telegraphy may be used for all station identification.

3AA-5.1 If you are using a frequency within a band designated to the amateur service on a secondary basis and another station assigned to a primary service on that band causes interference, what action should you take?

A. Notify the FCC's regional Engineer in Charge of the interference

B. Increase your transmitter's power to overcome the interference

C. Attempt to contact the station and request that it stop the interference

D. Change frequencies; you may also be causing interference to the other station and that would be a violation of FCC rules.

ANSWER D: Since our operation is on a secondary basis to the transmitting primary station, we must change frequencies to keep from causing interference to the other station.

3AA-5.2 What is the basic principle of frequency sharing between two stations allocated to a primary service within a frequency band, but each in a different ITU Region or Subregion?

    A. The station with a control operator holding a lesser class of license must yield the frequency to the station with a control operator holding a higher class license

    B. The station with a lower power output must yield the frequency to the station with a higher power output

    C. Both stations have an equal right to operate on the frequency

    D. Stations in ITU Regions 1 and 3 must yield the frequency to stations in ITU Region 2

ANSWER C: Hams must share the amateur frequencies. No ham owns a specific spot on the dial! All hams have an equal right to operate on any frequency that their class of license authorizes.

3AA-6-1.1 FCC Rules specify the maximum transmitter power that you may use with your amateur station. At what point in your station is the transmitter power measured?

    A. By measuring the final amplifier supply voltage inside the transmitter or amplifier

    B. By measuring the final amplifier supply current inside the transmitter or amplifier

    C. At the antenna terminals of the transmitter or amplifier

    D. On the antenna itself, after the feed line

ANSWER C: Put your watt meter into the antenna output of your transceiver. This is where you measure power output.

3AA-6-1.2 What is the term used to define the average power supplied to the antenna transmission line during one RF cycle at the crest of the modulation envelope?

    A. Peak transmitter power

    B. Peak output power

    C. Average radio-frequency power

    D. Peak envelope power

ANSWER D: If we measure to the crest of the modulation envelope, this is at its peak, and is called peak envelope power.

3AA-6-2.1 Notwithstanding the numerical limitations in the FCC Rules, how much transmitting power shall be used by an amateur station?

    A. There is no regulation other than the numerical limits.

    B. The minimum power level required to achieve S9 signal reports

    C. The minimum power necessary to carry out the desired communication

    D. The maximum power available, as long as it is under the allowable limit

ANSWER C: Always run as little power as possible to maintain communications. This also helps prevent interfering with your neighbor's television reception.

3AA-6-3.1 What is the maximum transmitting power permitted an amateur station on 146.52 MHz?

    A. 200 watts PEP output

    B. 500 watts ERP

C. 1000 watts dc input

D. 1500 watts PEP output

ANSWER D: When you upgrade to Technician, your new privileges will incorporate the popular 2-meter band. Would you believe you can run up to 1-1/2 kilowatts (1500 watts) on 2 meters? This is perfectly legal, but used only for those communications that need high power. This would include moon bounce, long-haul tropospheric ducting, Sporadic-E contacts, and meteor shower contacts.

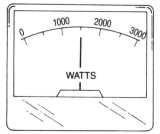

Maximum PEP Output on 2 Meters

3AA-6-4.1 What is the maximum transmitting power permitted an amateur station in beacon operation?

A. 10 watts PEP output

B. 100 watts PEP output

C. 500 watts PEP output

D. 1500 watts PEP output

ANSWER B: Radio beacons are employed by amateur operators to automatically transmit a distinctive call sign on a specific beacon band frequency. This allows other hams to tune them in from far away to determine band conditions. If you can hear a distant beacon thousands of miles away, chances are you can communicate on that Amateur Radio band to other stations thousands of miles away in the direction of the propagational beacon.

Sequence (Minutes)	Call Sign	Station Location
00	4U1UN/B	United Nations, New York City
01	W6WX/B	Stanford University, CA
02	KH60/B	Honolulu Community College, HI
03	JA2IGY	JARL, Mt. Asama, Japan
04	4X6TU/B	Tel Aviv University, Israel
05	OH2B	Helsinki Technical University, Finland
06	CT3B	ARRM (Madeira Radio Society), Madeira Island
07	ZS6DN/B	Transvaal, South Africa
08	LU4AA/B	Radio Club Argentino, Buenos Aires
09	Silent	

Each station in sequence transmits for one minute on 14.01 MHz.
The 09 minute, the last in the sequence, is silent; then the sequence repeats.

3AA-7-1.1 What is the maximum sending speed permitted for a RTTY transmission between 28 and 50 MHz?
A. 56 kilobauds
B. 19.6 kilobauds
C. 1200 bauds
D. 300 bauds

ANSWER C: The lower the frequency, the lower the baud rate allowed. Don't be disappointed—1200 bauds is pretty quick.

3AA-7-1.2 What is the maximum sending speed permitted for a RTTY transmission between 50 and 220 MHz?
A. 56 kilobauds
B. 19.6 kilobauds
C. 1200 bauds
D. 300 bauds

ANSWER B: The next possible answer up from 1200 bauds is 19.6 kilobauds (kilo means 1,000). 19,600 bauds is real quick!

3AA-7-1.3 What is the maximum sending speed permitted for a RTTY transmission above 220 MHz?
A. 300 bauds
B. 1200 bauds
C. 19.6 kilobauds
D. 56 kilobauds

ANSWER D: 56 kilobauds (56,000) is about the fastest that one would expect to go above 220 MHz.

3AA-7-2.1 What is the maximum frequency shift permitted for RTTY when transmitted below 50 MHz?
A. 100 Hz
B. 500 Hz
C. 1000 Hz
D. 5000 Hz

ANSWER C: When you listen to RTTY or F1B, you will hear two distinct tones, mark and space. The maximum separation allowed is 1,000 hertz, but 170 hertz is typical.

3AA-7-2.2 What is the maximum frequency shift permitted for RTTY when transmitted above 50 MHz?
A. 100 Hz or the sending speed, in bauds, whichever is greater
B. 500 Hz or the sending speed, in bauds, whichever is greater
C. The FCC rules do not specify a maximum frequency shift above 50 MHz.
D. 5000 Hz or the sending speed, in bauds, whichever is greater

ANSWER C: There are no rules for a specific maximum frequency shift above 50 MHz.

3AA-7-3.1 What is the maximum authorized bandwidth permitted an amateur station transmission between 50 and 255 MHz using a non-standard code?
A. 20 kHz
B. 50 kHz

C. The total bandwidth shall not exceed that of a single-sideband emission

D. The total bandwidth shall not exceed 10 times that of a CW emission

ANSWER A: 20 kHz is the authorized bandwidth of a signal within this range.

3AA-7-3.2 What is the maximum authorized bandwidth of a RTTY, data or multiplexed emission using an unspecified digital code within the frequency range of 220 to 450 MHz?

A. 50 kHz

B. 150 kHz

C. 200 kHz

D. 100 kHz

ANSWER D: 100 kHz is the maximum limit here.

3AA-7-3.3 What is the maximum authorized bandwidth of an RTTY, data or multiplexed emission using an unspecified digital code within the 420 to 450-MHz amateur band?

A. 50 kHz

B. 200 kHz

C. 300 kHz

D. 100 kHz

ANSWER D: Between 420 to 450 MHz, bandwidth is allowed up to 100 kHz.

3AA-8-1.1 How must a control operator who has a Novice license and a Certificate of Successful Completion of Examination for Technician privileges identify his/her station when transmitting on 146.34 MHz?

A. The new Technician may not operate on 146.34 until his or her new license arrives.

B. The licensee gives his or her call sign, followed by any suitable word that denotes the slant mark and the identifier "KT"

C. No special form of identification is needed.

D. The licensee gives his or her call sign and states the location of the VE examination where he or she obtained the certificate of successful completion.

ANSWER B: You can go on the air immediately after you pass your 25 multiple-choice question examination. You must use a special designator after your Novice call sign to identify your recent upgrade. It takes about 60 days for your new license to arrive. You may elect to keep your same entry-level call sign, or trade it in for a slightly shorter Technician/General Class call. When your new call sign arrives, you won't need to identify with your temporary identifier code anymore.

3AA-8-2.1 Which language(s) must be used when making the station identification by telephony?

A. The language being used for the contact may be used if it is not English, providing the US has a third-party traffic agreement with that country.

B. English must be used for identification.

C. Any language may be used, if the country which uses that language is a member of the International Telecommunication Union.

D. The language being used for the contact must be used for identification purposes.

ANSWER B: It's perfectly legal to speak to another station in a foreign language; however, when it comes time to identify, use English. You can also identify in Morse code.

3AA-8-3.1 What does the FCC recommend to aid correct station identification when using phone?
A. A speech compressor
B. Q signals
C. A recognized phonetic alphabet
D. Unique words of the operator's choice

ANSWER C: When talking with a new station that is unfamiliar with your call sign, use the phonetic alphabet. This is especially helpful when communicating with a foreign station that may not speak good English. Everyone uses the same phonetics, and it makes it easy to pick out someone's call sign.

A - Alpha	H - Hotel	O - Oscar	V - Victor
B - Bravo	I - India	P - Papa	W - Whiskey
C - Charlie	J - Juliette	Q - Quebec	X - X-ray
D - Delta	K - Kilo	R - Romeo	Y - Yankee
E - Echo	L - Lima	S - Sierra	Z - Zulu
F - Foxtrot	M - Mike	T - Tango	
G - Golf	N - November	U - Uniform	

Phonetic Alphabet

3AA-9-1.1 What is the term used to describe an amateur station transmitting communications for the purpose of observation of propagation and reception or other related experimental activities?
A. Beacon operation
B. Repeater operation
C. Auxiliary operation
D. Radio control operation

ANSWER A: You can tune in radio beacons near 14.1 MHz, and on 10 meters between 28.2 MHz and 28.3 MHz. They use CW to send their call signs over and over again for propagation phenomena information.

3AA-9-2.1 What class of amateur operator license must you hold to operate a beacon station?
A. Technician, General, Advanced or Amateur Extra class
B. General, Advanced or Amateur Extra class
C. Amateur Extra class only
D. Any license class

ANSWER A: Everyone but the Novice may operate a beacon station.

3AA-10.1 What is the maximum transmitter power an amateur station is permitted when transmitting signals to control a model craft?
A. One watt
B. One milliwatt
C. Two watts
D. Three watts

ANSWER A: If you ran more than one watt of power, every model in the country might take its command from your transmitter! One watt is a good power level to keep your model going within eyesight.

3AA-10.2 What minimum information must be indicated on the label affixed to a transmitter transmitting signals to control a model craft?
   A. Station call sign
   B. Station call sign and operating times
   C. Station call sign and the station licensee's name and address
   D. Station call sign, class of license, and operating times

ANSWER C: With your Technician Class license, you get to fly the coveted black flag. Just make sure you have all of your license information on the side of your transmitter.

3AA-10.3 What are the station identification requirements for an amateur station transmitting signals to control a model craft?
   A. Once every ten minutes, and at the beginning and end of each transmission
   B. Once every ten minutes
   C. At the beginning and end of each transmission
   D. Station identification is not required provided that a label indicating the station call sign and the station licensee's name and address is affixed to the station transmitter.

ANSWER D: Conceivably you could identify by letting your plane send your call sign in smoke signals and skywriting, but this is pretty crazy. No identification is required.

Skywriting Call Sign

3AA-10.4 Where must the writing indicating the station call sign and the licensee's name and address be affixed in order to operate under the special rules for radio control of remote model craft and vehicles?
   A. It must be in the operator's possession.
   B. It must be affixed to the transmitter.
   C. It must be affixed to the craft or vehicle.
   D. It must be filed with the nearest FCC Field Office.

ANSWER B: Just make sure your radio control transmitter has all of the proper license information affixed to it.

3AA-11-1.1 If an amateur repeater is causing harmful interference to another amateur repeater and a frequency coordinator has coordinated (recommends) the operation of one station and not the other, who is primarily responsible for resolving the interference?
  A. The licensee of the non-coordinated (unrecommended) repeater
  B. Both repeater licensees
  C. The licensee of the coordinated (recommended) repeater
  D. The frequency coordinator
ANSWER A: It's important to remember that all amateur repeaters must receive coordination. If you operate an amateur repeater that is uncoordinated, you are responsible for resolving interference to a coordinated repeater.

3AA-11-1.2 If an amateur repeater is causing harmful interference to another amateur repeater and a frequency coordinator has coordinated (recommends) the operation of both stations, who is primarily responsible for resolving the interference?
  A. The licensee of the repeater which has been coordinated for the longest period of time
  B. Both repeater licensees
  C. The licensee of the repeater which has been coordinated the most recently
  D. The frequency coordinator
ANSWER B: If both repeater stations are coordinated, both repeater licensees must mutually work out the interference problem.

3AA-11-1.3 If an amateur repeater is causing harmful interference to another amateur repeater and a frequency coordinator has not coordinated the operation of either station, who is primarily responsible for resolving the interference?
  A. Both repeater licensees
  B. The licensee of the repeater which has been in operation for the longest period of time
  C. The licensee of the repeater which has been in operation for the shortest period of time
  D. The frequency coordinator
ANSWER A: In this case, both repeaters are uncoordinated, so both licensees must resolve the problem.

3AA-11-2.1 Under what circumstances does the FCC declare a temporary state of communication emergency?
  A. When a declaration of war is received from Congress
  B. When the maximum usable frequency goes above 28 MHz
  C. When communications facilities in Washington, DC, are disrupted
  D. When a disaster disrupts normal communications systems in a particular area
ANSWER D: Ham radio operators always rise to the call for emergency communicators during widespread emergencies. Hurricanes, earthquakes, train derailments, airplane crashes, and widespread toxic spills are but a few of the emergencies that hams handle. When it's a big one, the FCC (Federal Communications Commission) may declare a general state of communications emergency. This allows the FCC to specifically regulate certain frequencies, and certain

designated areas where only emergency communications may take place within that area. Hams heading up the emergency communications will spread the word that the FCC has given them first priority to the frequency in a specific area.

3AA-11-2.2 By what means should a request for a declaration of a temporary state of communication emergency be initiated?
A. Communication with the FCC Engineer-In-Charge of the affected area
B. Communication with the US senator or congressman for the area affected
C. Communication with the local Emergency Coordinator
D. Communication with the Chief of the FCC Private Radio Bureau

ANSWER A: Amateur operators who are trained in emergency communications should first assess the emergency scene. If they need help in clearing a frequency, or controlling a designated area, the emergency communicator should seek the assistance of the local FCC engineer in charge. The FCC may then act to declare a general state of emergency for that area or for certain frequencies.

3AA-11-2.3 What information is included in an FCC declaration of a temporary state of communication emergency?
A. Designation of the areas affected and of organizations authorized to use radio communications in the affected area
B. Designation of amateur frequency bands for use only by amateurs participating in emergency communications in the affected area, and complete suspension of Novice operating privileges for the duration of the emergency
C. Any special conditions and special rules to be observed by stations during the communication emergency
D. Suspension of amateur rules regarding station identification and business communication

ANSWER C: If you are asked to stop transmitting on a certain frequency because it is reserved only for emergency communications, then by all means comply! Do listen in to see if there is anything that you might do to help—but avoid transmitting on the frequency unless directed to do so by the emergency net controller. If they are not asking for outside help, then don't transmit an offer for assistance.

3AA-11-2.4 If a disaster disrupts normal communication systems in an area where the amateur service is regulated by the FCC, what kinds of transmissions are authorized to amateur stations in such an area?
A. Communications which are necessary to meet essential communication needs and facilitate relief actions
B. Communications which allow a commercial business to continue to operate in the affected area
C. Communications for which material compensation has been paid to the amateur operator for delivery into the affected area
D. Communications which are to be used for program production or newsgathering for broadcasting purposed

ANSWER A: If you do take part in emergency communications, keep your transmissions as short as possible. Listen to airline pilots communications over the airwaves—you should adopt their brief style when taking part in emergency communications.

3AA-12.1 What is meant by the term *broadcasting?*
A. Transmissions intended for reception by the general public, either direct or relayed
B. Retransmission by automatic means of programs or signals emanating from any class of station other than amateur
C. The transmission of any one-way radio communication, regardless of purpose or content
D. Any one-way or two-way radio communication involving more than two stations

ANSWER A: You may not operate your station like an AM or shortwave broadcast station. You cannot transmit to the public directly.

3AA-12.2 Which of the following is an amateur station that cannot automatically retransmit signals of other amateur stations?
A. Auxiliary station
B. Repeater station
C. Beacon station
D. Space station

ANSWER C: Beacon stations are automatic transmitters without a receiver. You will find beacons on 14.1 MHz plus 28.2 to 28.3 MHz. Beacons are not capable of automatically retransmitting any other signals. All beacons do is transmit.

3AA-12.3 Which of the following is an amateur station that is permitted to automatically retransmit signals of other amateur stations?
A. Beacon station
B. Space station
C. Official bulletin station
D. RACES station

ANSWER B: The new 1990 decade of space stations may now relay, automatically, certain signals on amateur bands. The new microsats automatically retransmit packet information all over the world. Stations that do not automatically retransmit signals would be beacon, bulletin stations, and a RACES station.

3AA-12.4 Signals from what type of radio stations may be directly retransmitted by an amateur station?
A. AM radio station
B. Police or fire department radio station
C. NOAA weather station
D. US Government communications between the space shuttle and associated Earth stations with prior approval from the National Aeronautics and Space Administation (NASA)

ANSWER D: Regularly you will hear the excitement of NASA in space. Local repeater stations will rebroadcast the space shuttle communications. We are not allowed to automatically retransmit music AM stations, police stations, or even the weather stations.

3AA–COMMISSION'S RULES **3**

3AA-12.5 When may US Government communications between the space shuttle and associated Earth stations be directly retransmitted by an amateur station?

A. After prior approval has been obtained from the FCC in Washington, DC

B. No radio stations other than amateur may be retransmitted in the amateur service

C. After prior approval has been obtained from the National Aeronautics and Space Administration (NASA)

D. After prior approval has been obtained from the nearest FCC Engineer-In-Charge

ANSWER C: For a repeater operator to retransmit NASA information, they must first receive prior approval from the National Aeronautics & Space Administration.

3AA-13.1 What kinds of *one-way* communications by amateur stations are not considered broadcasting?

A. All types of one-way communications by Amateurs are considered by the FCC as broadcasting.

B. Beacon operation, remote control of a device, emergency communications, information bulletins consisting solely of subject matter of direct interest to the amateur service, and telegraphy practice

C. Only code-practice transmissions conducted simultaneously on all available amateur bands below 30 MHz and conducted for more than 40 hours per week are not considered broadcasting.

D. Only actual emergency communications during a declared communications emergency are exempt.

ANSWER B: Hams communicate with other hams. That's the whole idea behind the amateur service. However, radio amateurs may also take part in certain one-way communications where they don't expect, nor would hardly get, a response from the station receiving their transmissions.

3AA-13.2 Which of the following one-way communications may not be transmitted in the amateur service?

A. Transmissions to remotely control a device at a distant location

B. Transmissions to assist persons learning or improving their proficiency in Morse code

C. Brief transmissions to make adjustments to the station

D. Transmission of music

ANSWER D: Sorry, no singing or playing music on the air!

3AA-13.3 What kinds of one-way information bulletins may be transmitted by amateur stations?

A. NOAA weather bulletins

B. Commuter traffic reports from local radio stations

C. Regularly scheduled announcements concerning Amateur Radio equipment for sale or trade

D. Messages directed only to amateur operators consisting solely of subject matter of direct interest to the amateur service

133

ANSWER D: Every day Amateur Radio bulletin stations transmit the latest ham radio news. You won't hear the stock market, nor will you hear any news about the world—only ham radio news, and this is allowed.

3AA-13.4 What types of one-way amateur communications may be transmitted by an amateur station?

A. Beacon operation, radio control, code practice, retransmission of other services
B. Beacon operation, radio control, transmitting an unmodulated carrier, NOAA weather bulletins
C. Beacon operation, remote control of a device, information bulletins consisting solely of subject matter of direct interest to the amateur service, telegraphy practice and emergency communications
D. Beacon operation, emergency-drill-practice transmissions, automatic retransmission of NOAA weather transmissions, code practice

ANSWER C: The most well-known Amateur Radio bulletin station is operated by the non-profit American Radio Relay League. Bulletin station W1AW from Newington, Connecticut transmits on worldwide and local frequencies.

3AA-14.1 What types of material compensation, if any, may be involved in third-party traffic transmitted by an amateur station?

A. Payment of an amount agreed upon by the amateur operator and the parties involved
B. Assistance in maintenance of auxiliary station equipment
C. Donation of amateur equipment to the control operator
D. No compensation may be accepted.

ANSWER D: You cannot accept any type of payment for your ham radio operation. You cannot charge for handling messages on ham radio. You cannot accept a gift for handling ham radio traffic.

3AA-14.2 What types of business communications, if any, may be transmitted by an amateur station on behalf of a third party?

A. The FCC rules specifically prohibit communications with a business for any reason.
B. Business communications involving the sale of Amateur Radio equipment
C. Communications to a business may be provided during an emergency as provided by the FCC rules.
D. Business communications aiding a broadcast station

ANSWER C: This is a confusing question and answer. Except in an emergency, business communications are not allowed on ham radio. However, during an emergency, it would be perfectly legal to activate the phone patch and order 100 pizzas from the local pizzaria for the firefighters to eat while actually engaged in putting out a forest fire threatening nearby homes. Many times the American Red Cross will work closely with amateur operators with this type of traffic.

3AA-14.3 Does the FCC allow third-party messages when communicating with amateur operators in a foreign country?

A. Third-party messages with a foreign country are only allowed on behalf of other amateurs.
B. Yes, provided the third-party message involves the immediate family of one of the communicating amateurs

C. Under no circumstances may US amateurs exchange third-party messages with an amateur in a foreign country.

D. Yes, when communicating with a person in a country with which the US shares a third-party agreement

ANSWER D: Your ham radio station is not a substitute for the regular international telephone service. If a third party wishes to use your ham station to talk with another ham in a foreign country (with which there is a third-party agreement), the third party must keep the communications to a personal character of relative unimportance.

3AA-15.1 Under what circumstances, if any, may a third party participate in radio communications from an amateur station if the third party is ineligible to be a control operator of one of the stations?

A. A control operator must be present at the control point and continuously monitor and supervise the third party participation. Also, contacts may only be made with amateurs in the US and countries with which the US has a third-party communications agreement

B. A control operator must be present and continuously monitor and supervise the radio communication to ensure compliance with the rules only if contacts are made with amateurs in countries with which the US has no third-party traffic agreement.

C. A control operator must be present and continuously monitor and supervise the radio communication to ensure compliance with the rules. In addition, the control operator must key the transmitter and make the station identification.

D. A control operator must be present and continuously monitor and supervise the radio communication to ensure compliance with the rules. Also if contacts are made on frequencies below 30 MHz, the control operator must transmit the call signs of both stations.

ANSWER A: It's perfectly legal to let a non-licensed ham talk over your ham radio setup. This is a great way to demonstrate what ham radio is all about. You must stay at the mike at all times, and you should give your own call sign when it's time to identify. Don't let the third party give your call sign or have any control over your equipment. You are the ham, and you must be in control at all times.

3AA-15.2 Where must the control operator be situated when a third party is participating in radio communications from an amateur station?

A. If a radio remote control is used, the control operator may be physically separated from the control point, when provisions are incorporated to shut off the transmitter by remote control.

B. If the control operator supervises the third party until he or she is satisfied of the competence of the third party, the control operator may leave the control point.

C. The control operator must be present at the control point

D. If the third party holds a valid radiotelegraph license issued by the FCC, no supervision is necessary.

ANSWER C: Don't even leave the room when a third party is using your ham set. Your license is at stake, so stay right there with the third party to insure compliance with all FCC rules.

See pg. 213 in Appendix for a list of third-party countries.

**3AA-15.3 What must the control operator do while a third party is partici-
pating in radio communications?**
   A. If the third party holds a valid commercial radiotelegraph license, no
      supervision is necessary.
   B. The control operator must tune up and down 5 kHz from the trans-
      mitting frequency on another receiver, to ensure that no interference
      is taking place.
   C. If a radio control link is available, the control operator may leave the
      room.
   D. The control operator must continuously monitor and supervise the
      third party's participation.
ANSWER D: If the third party is speaking a foreign language, you must know
exactly what they are saying. Also, make sure there is a third-party agreement
with that foreign amateur before allowing your third party to go on the air.

**3AA-15.4 In an exchange of international third-party communications,
when is the station identification procedure required?**
   A. Only at the beginning of the communications
   B. At the end of each exchange of communications
   C. The station identification procedure is not required during interna-
      tional third-party communications
   D. Only at the end of multiple exchanges of communications
ANSWER B: Make certain that the station licensee gives the official call sign at the
end of each exchange of communication during third-party communications. Do
not allow the third party to actually give the station I.D.

**3AA-16.1 Under what circumstances, if any, may an amateur station
transmit radio communications containing obscene words?**
   A. Obscene words are permitted when they do not cause interference to
      any other radio communication or signal.
   B. Obscene words are prohibited in Amateur Radio transmissions.
   C. Obscene words are permitted when they are not retransmitted
      through repeater or auxiliary stations.
   D. Obscene words are permitted, but there is an unwritten rule among
      amateurs that they should not be used on the air.
ANSWER B: No ham operator likes talking to a potty mouth. Don't use expletives
over the airwaves.

**3AA-16.2 Under what circumstances, if any, may an amateur station
transmit radio communications containing indecent words?**
   A. Indecent words are permitted when they do not cause interference to
      any other radio communication or signal.
   B. Indecent words are permitted when they are not retransmitted
      through repeater or auxiliary stations.
   C. Indecent words are permitted, but there is an unwritten rule among
      Amateurs that they should not be used on the air.
   D. Indecent words are prohibited in Amateur Radio transmissions.
ANSWER D: Any type of indecent language is frowned upon by other hams.
Remember, ham radio is a family hobby, and there may be young kids out there
listening.

3AB–OPERATING PROCEDURES **3**

3AA-16.3 Under what circumstances, if any, may an amateur station transmit radio communications containing profane words?
   A. Profane words are permitted when they are not retransmitted through repeater or auxiliary stations.
   B. Profane words are permitted, but there is an unwritten rule among Amateurs that they should not be used on the air.
   C. Profane words are prohibited in Amateur Radio transmissions.
   D. Profane words are permitted when they do not cause interference to any other radio communication or signal.

ANSWER C: Using profane language is just not necessary on the ham bands. Watch your tongue, and communicate like a professional.

3AA-17.1 Which of the following VHF/UHF bands may not be used by Earth stations for satellite communications?
   A. 6 meters
   B. 2 meters
   C. 1.25 meters
   D. 70 centimeters

ANSWER A: There are no uplink frequencies found on the 6-meter band for satellite communications.

Subelement 3AB – Operating Procedures (3 examination questions from 30 questions in 3AB)

3AB-1.1 What is the meaning of: "Your report is five seven..."?
   A. Your signal is perfectly readable and moderately strong.
   B. Your signal is perfectly readable, but weak.
   C. Your signal is readable with considerable difficulty.
   D. Your signal is perfectly readable with near pure tone.

ANSWER A: Looking at the RST signal report system chart, a 5 by 7 report is a pretty good indication that the other station is hearing you well. Since there is no third number, the report is for a voice transmission.

3AB-1.2 What is the meaning of: "Your report is three three..."?
   A. The contact is serial number thirty-three.
   B. The station is located at latitude 33 degrees.
   C. Your signal is readable with considerable difficulty and weak in strength.
   D. Your signal is unreadable, very weak in strength.

ANSWER C: Since only two numbers are given, you must be using voice. Better check out your equipment—a 3 by 3 report is not very good!

RST Reporting Chart
Source: ARRL

The following codes are used in the RST System:

Readability
1 - Unreadable
2 - Barely readable; occasional words distinguishable.
3 - Readable with considerable difficulty
4 - Readable with practically no difficulty
5 - Perfectly readable.

Signal Strength
1 - Faint and barely perceptible signals
2 - Very weak signals
3 - Weak signals
4 - Fair signals
5 - Fairly good signals
6 - Good signals
7 - Moderately strong signals
8 - Strong signals
9 - Extremely strong signals

Tone
1 - Very rough, broad signals, 60 cycle AC may be present
2 - Very rough AC tone, harsh, broad
3 - Rough, low pitched AC tone, no filtering
4 - Rather rough AC tone, some trace of filtering
5 - Filtered rectified AC note, musical, ripple modulated
6 - Slight trace of filtered tone but with ripple modulation
7 - Near DC tone but trace of ripple modulation
8 - Good DC tone, may have slight trace of modulation
9 - Purest, perfect DC tone with no trace of ripple or modulation.

* The TONE report refers only to the purity of the signal, and has no connection with its stability or freedom from clicks or chirps. If the signal has the characteristic steadiness of crystal control, add X to the report (e.g., RST 469X). If it has a chirp or "tail" (either on "make" or "break") add C (e.g., RST 469C). If it has clicks or other noticeable keying transients, add K (e.g., RST 469K). If a signal has both chirps and clicks, add both C and K (e.g., RST 469CK).

3AB-1.3 What is the meaning of: "Your report is five nine plus 20 dB..."?
 A. Your signal strength has increased by a factor of 100.
 B. Repeat your transmission on a frequency 20 kHz higher.
 C. The bandwidth of your signal is 20 decibels above linearity.
 D. A relative signal-strength meter reading is 20 decibels greater than strength 9.

ANSWER D: Any signal over S9 is an excellent one. Most worldwide sets have well-calibrated S-meters that register 10, 20, 40, and 60 dB over S9. Your signal is plenty strong! Although the question and answer is worded technically correct, most hams would simply state that "your signal is 20 over 9". Same thing, but less formal on the air.

3AB-2-1.1 How should a QSO be initiated through a station in repeater operation?
 A. Say "breaker, breaker 79."
 B. Call the desired station and then identify your own station.
 C. Call "CQ" three times and identify three times.
 D. Wait for a "CQ" to be called and then answer it.

ANSWER B: When we QSO with a station, we communicate with it. That's what QSO means. If the repeater is available, simply call the other station by its call sign, and then give your own call sign.

3AB-2-1.2 Why should users of a station in repeater operation pause briefly between transmissions?
    A. To check the SWR of the repeater
    B. To reach for pencil and paper for third party traffic
    C. To listen for any hams wanting to break in
    D. To dial up the repeater's autopatch
ANSWER C: A repeater is like a party line—there may be others who may wish to use the system. In an emergency, stations may break in saying "Break, Break, Break". Give up the channel immediately. Always leave enough time between picking up the conversation for other stations to break in. It's a pause that may refresh someone else's day in an emergency.

3AB-2-1.3 Why should users of a station in repeater operation keep their transmissions short and thoughtful?
    A. A long transmission may prevent someone with an emergency from using the repeater.
    B. To see if the receiving station operator is still awake
    C. To give any non-hams that are listening a chance to respond
    D. To keep long-distance charges down
ANSWER A: During peak traffic hours, keep your transmissions short. Repeaters are a great way to find out traffic reports and for reporting traffic accidents.

3AB-2-1.4 What is the proper procedure to break into an on-going QSO through a station in repeater operation?
    A. Wait for the end of a transmission and start calling.
    B. Shout, "break, break!" to show that you're eager to join the conversation.
    C. Turn on your 100-watt amplifier and override whoever is talking.
    D. Send your call sign during a break between transmissions.
ANSWER D: A good way to join an ongoing repeater conversation is to quickly drop the last two letters of your call sign in between the transmissions. Remember, other stations will always leave a pause to allow stations to join in on the QSO.

3AB-2-1.5 What is the purpose of repeater operation?
    A. To cut your power bill by using someone's higher power system
    B. To enable mobile and low-power stations to extend their usable range
    C. To reduce your telephone bill
    D. To call the ham radio distributor 50 miles away
ANSWER B: Repeaters are sponsored by ham radio clubs and individual hams for everyone to use. They are usually placed high atop a mountain or a very tall building. Mobile and portable sets operate through repeaters with dramatically extended range. Even base stations are permitted to use repeaters for added communications distance. There is usually no charge for joining a repeater group. Some repeaters have autopatch, and those repeaters may require special access codes and financial support.

3AB-2-1.6 What is meant by "making the repeater time out"?
   A. The repeater's battery supply has run out.
   B. The repeater's transmission time limit has expired during a single transmission.
   C. The warranty on the repeater duplexer has expired.
   D. The repeater is in need of repairs.

ANSWER B: If you are long-winded, you could time out a repeater. Most repeaters have a one minute timer to limit any one transmission. If you out-talk the timer, the repeater shuts off, and won't come back on for about a minute or so. If you must really be long-winded, allow the repeater to reset by momentarily dropping your carrier until it goes "beep".

3AB-2-1.7 During commuting rush hours, which types of operation should relinquish the use of the repeater?
   A. Mobile operators
   B. Low-power stations
   C. Highway traffic information nets
   D. Third-party communications nets

ANSWER D: It's not a good idea to let a friend talk over your microphone as a third party during heavy repeater use time. During rush hours, most repeaters are used for traffic advisories and traffic accident reports.

3AB-2-2.1 Why should simplex be used where possible instead of using a station in repeater operation?
   A. Farther distances can be reached
   B. To avoid long distance toll charges
   C. To avoid tying up the repeater unnecessarily
   D. To permit the testing of the effectiveness of your antenna

ANSWER C: If you are close to the station you are communicating with, switch over to a direct channel. This is called simplex. Every band has several simplex channels specifically designed to relieve repeater congestion. You might be surprised how far simplex will go.

3AB-2-2.2 When a frequency conflict arises between a simplex operation and a repeater operation, why does good amateur practice call for the simplex operation to move to another frequency?
   A. The repeater's output power can be turned up to ruin the front end of the station in simplex operation.
   B. There are more repeaters than simplex operators.
   C. Changing the repeater's frequency is not practical.
   D. Changing a repeater frequency requires the authorization of the Federal Communications Commission.

ANSWER C: Sometimes you might accidentally operate on a repeater input or output channel without realizing you are on a repeater pair. If you are talking to another local station on simplex, and are informed you're on a repeater duplex channel, you are obliged to change frequency. This makes sense—all you need to do is to spin the dial. Repeaters can't change frequency because they are coordinated, so the burden is on you to move to an authorized simplex channel. You could also move to a repeater input or output that has no system on the air in your area.

3AB-2-3.1 What is the usual input/output frequency separation for stations in repeater operation in the 2-meter band?
  A. 1 MHz
  B. 1.6 MHz
  C. 170 Hz
  D. 0.6 MHz
ANSWER D: You will have this question, or one of the next three questions, on your exam, for sure. These are important to remember, too. Although some sets already come with the repeater splits memorized, some sets require initializing. On the 2-meter wavelength band, repeater inputs and outputs are usually separated by 600 kHz, which is the same thing as 0.6 MHz on your examination.

3AB-2-3.2 What is the usual input/output frequency separation for stations in repeater operation in the 70-centimeter band?
  A. 1.6 MHz
  B. 5 MHz
  C. 600 kHz
  D. 5 kHz
ANSWER B: 70 centimeters (0.70 meters) is the 450-MHz band, and input and output repeater separation is 5 MHz.

3AB-2-3.3 What is the usual input/output frequency separation for a 6-meter station in repeater operation?
  A. 1 MHz
  B. 600 kHz
  C. 1.6 MHz
  D. 20 kHz
ANSWER A: Although there is not much 6-meter repeater activity, input and output repeater separation is 1 MHz.

3AB-2-3.4 What is the usual input/output frequency separation for a 1.25-meter station in repeater operation?
  A. 1000 kHz
  B. 600 kHz
  C. 1600 kHz
  D. 1.6 GHz
ANSWER C: The 1.25-meter wavelength band is 220 MHz, and repeater separation is 1.6 MHz, which is the same thing as 1600 kHz on your exam. Watch out for Answer D—it's gigahertz and it's not correct, but looks pretty close.

3AB-2-4.1 What is a repeater frequency coordinator?
  A. Someone who coordinates the assembly of a repeater station
  B. Someone who provides advice on what kind of system to buy
  C. The club's repeater trustee
  D. A person or group that recommends frequency pairs for repeater usage
ANSWER D: Hams can put up their own repeaters, but only with coordination from a local frequency repeater coordinating committee. These committees insure that any new repeater system will fit in with existing systems.

3AB-3.1 Why should local amateur communications be conducted on VHF and UHF frequencies?
  A. To minimize interference on HF bands capable of long-distance sky-wave communication
  B. Because greater output power is permitted on VHF and UHF
  C. Because HF transmissions are not propagated locally
  D. Because absorption is greater at VHF and UHF frequencies

ANSWER A: We use VHF and UHF frequencies for operating on the 6-meter, 2-meter, 220-MHz, 450-MHz, and 1270-MHz FM ham bands. These bands are so high in frequency that they are line-of-sight to repeaters. We would use world-wide frequencies below 30 MHz for longer skywave range. If we wish to talk locally, we go to local VHF and UHF band frequencies.

3AB-3.2 How can on-the-air transmissions be minimized during a lengthy transmitter testing or loading up procedure?
  A. Choose an unoccupied frequency.
  B. Use a dummy antenna.
  C. Use a non-resonant antenna.
  D. Use a resonant antenna that requires no loading up procedure.

ANSWER B: If you work on your transmitter a lot, you may wish to test it without actually putting a signal out on the airwaves. A dummy antenna might be a lightbulb or a 50-ohm non-inductive resistance. The energy doesn't go very far. It's a great way to test your set before adding the antenna to it. It lets you run your rig at full output, but without actually going on the air.

3AB-3.3 What is the proper Q signal to use to determine whether a frequency is in use before making a transmission?
  A. QRV?
  B. QRU?
  C. QRL?
  D. QRZ?

ANSWER C: See appendix for Q codes. For the Technician Class exam, you don't need to memorize all of the international Q signals. But this one you need to know—QRL. Just think of the letter "L" in QRL as in "listening"—you listen before transmitting.

3AB-4.1 What is the proper distress calling procedure when using telephony?
  A. Transmit MAYDAY
  B. Transmit QRRR
  C. Transmit QRZ
  D. Transmit SOS

ANSWER A: Look at the words "telephony" and "telegraphy." They look almost the same in this and the following question, don't they? Don't speed read the exam. When using a telephone (telephony), say the word "Mayday".

VOICE
MAYDAY, MAYDAY, MAYDAY ⸺
CW
• • • ▬ ▬ ▬ • • •
S   O   S

International Distress Signals

3AB-4.2 What is the proper distress calling procedure when using telegraphy?
A. Transmit MAYDAY
B. Transmit QRRR
C. Transmit QRZ
D. Transmit SOS

ANSWER D: When using a telegraph key (telegraphy), transmit SOS "dit dit dit, dah dah dah, dit dit dit" for an emergency.

3AB-5-1.1 What is one requirement you must meet before you can participate in RACES drills?
A. You must be registered with ARRL.
B. You must be registered with a local racing organization.
C. You must be registered with the responsible civil defense organization.
D. You need not register with anyone to operate RACES.

ANSWER C: RACES stands for Radio Amateur Civil Emergency Service. It is a division of the Civil Defense organization. You must be registered to take part in RACES drills.

3AB-5-1.2 What is the maximum amount of time allowed per week for RACES drills?
A. Eight hours
B. One hour
C. As many hours as you want
D. Six hours, but not more than one hour per day

ANSWER B: No more than one hour may be allowed for RACES drills per week.

3AB-5-2.1 How must you identify messages sent during a RACES drill?
A. As emergency messages
B. As Amateur traffic
C. As official government messages
D. As drill or test messages

ANSWER D: To eliminate a misunderstanding of a drill message versus the real thing, always announce messages for practice as drill or test messages. You never know how many scanner monitor listeners are out there tuning in the ham bands!

3AB-6-1.1 What is the term used to describe first-response communications in an emergency situation?
A. Tactical communications
B. Emergency communications
C. Formal message traffic
D. National Traffic System messages

ANSWER A: Tactical communications are really exciting because they normally are occurring at the disaster scene.

3AB-6-1.2 What is one reason for using tactical call signs such as "command post" or "weather center" during an emergency?
A. They keep the general public informed about what is going on.
B. They promote efficiency and coordination in public-service communications activities.
C. They are required by the FCC.
D. They promote goodwill among amateurs.

ANSWER B: It's perfectly legal to use such tactical words as "command post," "triage team," or "disaster communicator" during an emergency. This promotes efficiency in the ham radio communications being provided.

3AB-6-2.1 What is the term used to describe messages sent into or out of a disaster area that pertain to a person's well being?
A. Emergency traffic
B. Tactical traffic
C. Formal message traffic
D. Health and welfare traffic

ANSWER D: This type of traffic deserves priority because we are talking about the welfare of human lives.

3AB-6-3.1 Why is it important to provide a means of operating your amateur station separate from the commercial ac power lines?
A. So that you can take your station mobile
B. So that you can provide communications in an emergency
C. So that you can operate field day
D. So that you will comply with Subpart 97.169 of the FCC Rules

ANSWER B: The author's station operates on solar panel power. The battery is safely outside, and the solar panels keep the battery charged even though he uses his ham station often.

3AB-6-3.2 Which type of antenna would be a good choice as part of a portable HF amateur station that could be set up in case of a communications emergency?
A. A three-element quad
B. A three-element Yagi
C. A dipole
D. A parabolic dish

ANSWER C: There is no simpler antenna than the common dipole. Remember the formula for constructing a dipole? Divide the frequency in MHz into 468 to obtain the end-to-end length in feet. Go back to your Novice questions and brush up on the equation used for the calculation.

Subelement 3AC – Radio-Wave Propagation (3 examination questions from 30 questions in 3AC)

3AC-1-1.1 What is the ionosphere?

A. That part of the upper atmosphere where enough ions and free electrons exist to affect radio-wave propagation

B. The boundary between two air masses of different temperature and humidity, along which radio waves can travel

C. The ball that goes on the top of a mobile whip antenna

D. That part of the atmosphere where weather takes place

ANSWER A: The ionosphere is the electrified atmosphere from 40 miles to 400 miles above the earth. You can sometimes see it as "northern lights." It is charged up daily by the sun, and does some miraculous things to radio waves that strike it. Some radio waves are absorbed, during daylight hours, by the ionosphere's D layer. Others are bounced back to earth. Yet others penetrate the ionosphere, and never come back again. The wavelength of the radio waves determines whether the waves will be absorbed, reflected, or penetrated. Here's a quick way to memorize what the different layers do during day and nighttime hours:The D layer is about 40 miles up. The D layer is a Daylight layer; it almost disappears at night. D for *Daylight*. The D layer absorbs radio waves between 1 MHz to 7 MHz. These are long wavelengths. All others pass through. The E layer is also a daylight layer, and it is very Eccentric. E for *Eccentric*. Patches of E layer ionization may cause some surprising reflections of signals on both high frequency as well as very high frequency. The E layer height is usually 70

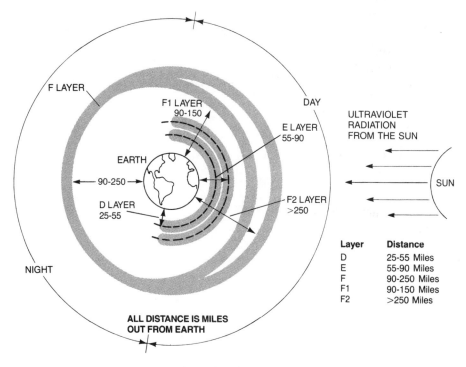

Ionosphere Layers

Source: *Antennas – Selection and Installation,* © 1986 Master Publishing, Inc., Richardson, Texas

miles.The F1 layer is one of the furthest layers away. The F layer gives us those Far away signals. F for *Far away*. The F1 layer is present during daylight hours, and is up around 150 miles. The F2 layer is also present during daylight hours, and it gives us the Furthest range. The F2 layer is 250 miles high, and it's the best for the *Furthest* range on medium and short waves. At nighttime, the F1 and F2 layers combine to become just the F layer at 180 miles. This F layer at nighttime will usually bend radio waves between 1 MHz and 15 MHz back to earth. At night, the D and E layers disappear.

3AC-1-1.2 What is the region of the outer atmosphere that makes long-distance radio communications possible as a result of bending of radio waves?
  A. Troposphere
  B. Stratosphere
  C. Magnetosphere
  D. Ionosphere
ANSWER D: When radio waves in the medium and high frequency range strike the ionosphere, they are often reflected, refracted, or simply bent back to earth. This occurs in the ionosphere.

3AC-1-1.3 What type of solar radiation is most responsible for ionization in the outer atmosphere?
  A. Thermal
  B. Ionized particle
  C. Ultraviolet
  D. Microwave
ANSWER C: It's the ultraviolet component of the sun's radiation that creates our ionosphere. More ultraviolet radiation on any one day will lead us to either improved or disturbed radio conditions.

3AC-1-2.1 Which ionospheric layer limits daytime radio communications in the 80-meter band to short distances?
  A. D layer
  B. F1 layer
  C. E layer
  D. F2 layer
ANSWER A: The 80-meter wavelength band is around 3.75 MHz. This is absorbed by the D layer during daylight conditions.

3AC-1-2.2 What is the lowest ionospheric layer?
  A. The A layer
  B. The D layer
  C. The E layer
  D. The F layer
ANSWER B: Be careful on this question—they simply ask for the lowest ionospheric layer, not necessarily the lowest one that gives us skip. Since the D layer is the lowest, this is the correct answer. See figure at question 3AC-1-1.1.

3AC-1-3.1 What is the lowest region of the ionosphere that is useful for long-distance radio wave propagation?
  A. The D layer
  B. The E layer

C. The F1 layer
D. The F2 layer

ANSWER B: Since the D layer only absorbs, we cannot choose it as the correct answer for a region that causes signals to bounce back to earth. The first layer up that gives us bounce capabilities is that eccentric E layer. It's only present during daylight hours.

3AC-1-4.1 Which layer of the ionosphere is mainly responsible for long-distance sky-wave radio communications?
A. D layer
B. E layer
C. F1 layer
D. F2 layer

ANSWER D: Since the F2 layer is higher than the F1 layer, it would be mainly responsible for the longest distance skywave hop back to earth. See figure at question 3AC-1-1.1.

3AC-1-4.2 What are the two distinct sub-layers of the F layer of the ionosphere during the daytime?
A. Troposphere and stratosphere
B. F1 and F2
C. Electrostatic and electromagnetic
D. D and E

ANSWER B: During daylight hours, the sun is so powerful it actually breaks the F layer into two distinct regions—F1 and F2.

3AC-1-4.3 Which two daytime ionospheric layers combine into one layer at night?
A. E and F1
B. D and E
C. F1 and F2
D. E1 and E2

ANSWER C: Remember that F layer? It breaks apart into two distinct layers during the day, and recombines into simply the F layer at night.

3AC-2.1 Which layer of the ionosphere is most responsible for absorption of radio signals during daylight hours?
A. The E layer
B. The F1 layer
C. The F2 layer
D. The D layer

ANSWER D: The key word in this question is "absorption." Only one layer absorbs radio signals like a sponge, and that's the D layer during daylight hours. It absorbs medium and low frequency signals only. Higher frequencies pass through the D layer, and bounce off of other layers.

3AC-2.2 When is ionospheric absorption most pronounced?
A. When tropospheric ducting occurs
B. When radio waves enter the D layer at low angles
C. When radio waves travel to the F layer
D. When a temperature inversion occurs

ANSWER B: We have most absorption to long wavelength signals. Long wavelength signals enter the D layer at low angles.Remember: "Lower-Longer." We'll use this memory jogger again for antenna length.

3AC-2.3 During daylight hours, what effect does the D layer of the ionosphere have on 80-meter radio waves?
 A. The D layer absorbs the signals.
 B. The D layer bends the radio waves out into space.
 C. The D layer refracts the radio waves back to earth.
 D. The D layer has little or no effect on 80 meter radio wave propagation.
ANSWER A: Here's that 80-meter wavelength band again during daylight hours—it's almost completely absorbed by the D layer.

3AC-2.4 What causes ionospheric absorption of radio waves?
 A. A lack of D layer ionization
 B. D layer ionization
 C. The presence of ionized clouds in the E layer
 D. Splitting of the F layer
ANSWER B: Just think of the D layer as that "Darn sponge" that absorbs long wavelength signals during daylight hours. Absorption always takes place in the D layer during daylight.

3AC-3.1 What is usually the condition of the ionosphere just before sunrise?
 A. Atmospheric attenuation is at a maximum.
 B. Ionization is at a maximum.
 C. The E layer is above the F layer.
 D. Ionization is at a minimum.
ANSWER D: Just as the outside temperature is at it's coldest, the amount of ionization left in the ionosphere just before sunrise is at a minimum.

3AC-3.2 At what time of day does maximum ionization of the ionosphere occur?
 A. Dusk
 B. Midnight
 C. Midday
 D. Dawn
ANSWER C: Just as the daily temperature is usually highest during midday, the maximum ionization occurs during midday.

3AC-3.3 Minimum ionization of the ionosphere occurs daily at what time?
 A. Shortly before dawn
 B. Just after noon
 C. Just after dusk
 D. Shortly before midnight
ANSWER A: Minimum ionization occurs just before the sun gets up. Maximum ionization occurs around midday.

3AC-3.4 When is E layer ionization at a maximum?
 A. Dawn
 B. Midday
 C. Dusk
 D. Midnight

ANSWER B: Here's that ionization question again, and again! During midday, it's maximum. Just before dawn, it's minimum.

3AC-4.1 What is the name for the highest radio frequency that will be refracted back to earth?
A. Lowest usable frequency
B. Optimum working frequency
C. Ultra high frequency
D. Critical frequency

ANSWER D: Any frequency that doesn't get refracted back to earth is usually higher than the "critical frequency." For long-range skywave propagation, operate on a ham band just below the critical frequency for best results.

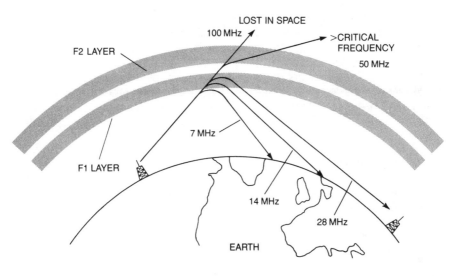

Critical Frequency

Source: *Antennas – Selection and Installation,* ©1986 Master Publishing, Inc., Richardson, Texas

3AC-4.2 What causes the maximum usable frequency to vary?
A. Variations in the temperature of the air at ionospheric levels
B. Upper-atmospheric wind patterns
C. The amount of ultraviolet and other types of radiation received from the sun
D. Presence of ducting

ANSWER C: Worldwide band conditions will vary from day to day. Some changes are predictable—such as seasonal changes and time of day changes. Others may be sudden due to intense solar activity on the sun creating changes in ultraviolet radiation down here on the earth. The sun's activity is important to monitor.

3AC-4.3 What does the term *maximum usable frequency* refer to?
A. The maximum frequency that allows a radio signal to reach its destination in a single hop
B. The minimum frequency that allows a radio signal to reach its destination in a single hop

C. The maximum frequency that allows a radio signal to be absorbed in the lowest ionospheric layer

D. The minimum frequency that allows a radio signal to be absorbed in the lowest ionospheric layer

ANSWER A: Maximum usable frequency is abbreviated "MUF." Monthly magazines predict the maximum usable frequency from here to there, anywhere in the world, at any time.

3AC-5.1 When two stations are within each other's skip zone on the frequency being used, what mode of propagation would it be desirable to use?

A. Ground wave propagation

B. Sky wave propagation

C. Scatter-mode propagation

D. Ionospheric ducting propagation

ANSWER C: It's quite possible your skywaves will actually skip over the station you wish to communicate with. You might switch to a lower band to bring in your range. Another trick is to turn on the power amplifier, and operate scatter mode communications. The extra amount of transmit power will bounce signals back to the normally vacant skip zone. Sometimes they bounce off to the side, and sometimes they bounce ahead. All of this is known as back scatter, side scatter, and forward scatter. If you have enough power and a directional antenna, scatter communications will help fill in the skip zone.

3AC-5.2 You are in contact with a distant station and are operating at a frequency close to the maximum usable frequency. If the received signals are weak and somewhat distorted, what type of propagation are you probably experiencing?

A. Tropospheric ducting

B. Line-of-sight propagation

C. Backscatter propagation

D. Waveguide propagation

ANSWER C: It's unbelievable, but sometimes you can talk to stations in South America with your antenna pointed west, and with better results in the evening hours.

3AC-6.1 What is the transmission path of a wave that travels directly from the transmitting antenna to the receiving antenna called?

A. Line of sight

B. The sky wave

C. The linear wave

D. The plane wave

ANSWER A: When we operate on VHF and UHF frequencies, our signals travel line-of-sight. They are too high in frequency to be reflected back by the ionosphere. The line-of-sight signals may travel some pretty strange routes to that distant station, too. Sometimes they are reflected by tall buildings. Sometimes they are bent over mountains, and sometimes they are funneled through hills. Yet regardless of the path, they are known as line-of-sight transmissions.

3AC-6.2 How are VHF signals within the range of the visible horizon propagated?

A. By sky wave
B. By direct wave
C. By plane wave
D. By geometric wave

ANSWER B: If you can see it, you can work it on VHF and UHF via direct wave.

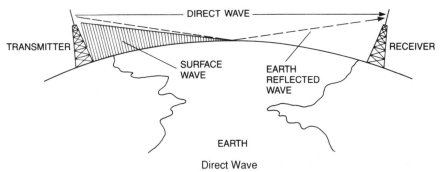

Direct Wave

Source: *Antennas – Selection and Installation*, ©1986 Master Publishing, Inc., Richardson, Texas

3AC-7.1 Ducting occurs in which region of the atmosphere?

A. F2
B. Ionosphere
C. Troposphere
D. Stratosphere

ANSWER C: The key word in this question is "ducting." In your home you use ducts to shuffle that warm or cool air around the house. Out in radio land, natural atmospheric ducts form that shuffle VHF and UHF radio waves well beyond line-of-sight range. Tropospheric ducting occurs most often during the summer months, and sometimes occurs in the presence of large storm systems. Gordon West, author of this book, is one of the record holders in tropospheric ducting on VHF line-of-sight frequencies between his home near Los Angeles all the way over to Hawaii. This is not skip off the ionosphere, but rather tropospheric ducting several hundred feet above the water all those thousands of miles away!

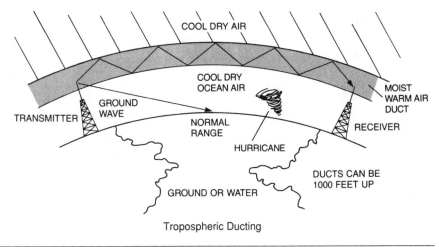

Tropospheric Ducting

3AC-7.2 What effect does tropospheric bending have on 2-meter radio waves?

A. It increases the distance over which they can be transmitted.
B. It decreases the distance over which they can be transmitted.
C. It tends to garble 2-meter phone transmissions.
D. It reverses the sideband of 2-meter phone transmissions.

ANSWER A: The nice thing about tropospheric bending and ducting is it gives us some extraordinary range that we normally don't get under average weather conditions.

3AC-7.3 What atmospheric phenomenon causes tropospheric ducting of radio waves?

A. A very low pressure area
B. An aurora to the north
C. Lightning between the transmitting and receiving station
D. A temperature inversion

ANSWER D: Have you ever seen a mirage? Out on the desert it looks like blue water instead of sand ahead. Actually that's the blue sky you are looking at. Light waves that normally travel in straight lines bounce off the super-heated wind-less sand and pavement and are reflected back to your eyes. Same thing during a trophospheric duct—but just backwards. Typically straight line VHF and UHF signals begin to travel up and away, but are bent back by a sharp boundary layer of warm, moist air overlying cool, dry air below and above. In the city, this is what traps the smog and gives us one of those unbearable days. Get on the radio—it will be unbelievable. See figure at question 3AC-7.1.

3AC-7.4 Tropospheric ducting occurs as a result of what phenomenon?

A. A temperature inversion
B. Sun spots
C. An aurora to the north
D. Lightning between the transmitting and receiving station

ANSWER A: You can watch your local weather maps for signs of temperature inversions. If they predict bad air quality, you can predict good band qualities on VHF and UHF.

3AC-7.5 What atmospheric phenomenon causes VHF radio waves to be propagated several hundred miles through stable air masses over oceans?

A. Presence of a maritime polar air mass
B. A widespread temperature inversion
C. An overcast of cirriform clouds
D. Atmospheric pressure of roughly 29 inches of mercury or higher

ANSWER B: Some of these tropospheric ducts may extend out to 1,000 miles. The record is between Gordon West's (your author) home in California and Hawaii. Any hams in the area are invited to stop by and see his record breaking tropo station.

3AC-7.6 In what frequency range does tropospheric ducting occur most often?

A. LF
B. MF

C. HF

D. VHF

ANSWER D: The 2-meter wavelength band is one of the best for taking advantage of tropospheric ducting conditions. With your new Technician Class license around the corner, start planning for your 2-meter station now.

Subelement 3AD – Amateur Radio Practice (4 examination questions from 44 questions in 3AD)

3AD-1-1.1 Where should the green wire in an ac line cord be attached in a power supply?

A. To the fuse

B. To the "hot" side of the power switch

C. To the chassis

D. To the meter

ANSWER C: Green is ground. That's easy to remember. The chassis of our radio equipment is the metal top, bottom, and side covers that keep everything together on the inside. Most radio sets offer a ground connection on their rear. A green ground connection conductor will let you keep your colors straight.

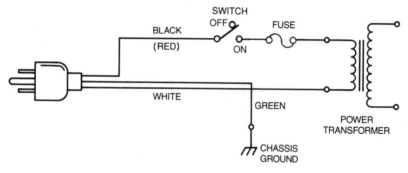

AC Line Connections

3AD-1-1.2 Where should the *black (or red)* wire in a three-wire line cord be attached in a power supply?

A. To the filter capacitor

B. To the dc ground

C. To the chassis

D. To the fuse

ANSWER D: We always fuse the hot lead. Most ac line cords use a black— sometimes red—wire in the three-wire cord that plugs into the wall socket. This goes to the fuse. It's hot, so don't touch it when everything's plugged in. See figure at question 3AD-1-1.1.

3AD-1-1.3 Where should the *white wire* in a three-wire line cord be attached in a power supply?

A. To the side of the transformer's primary winding that has a fuse

B. To the side of the transformer's primary winding without a fuse

C. To the black wire

D. To the rectifier junction

ANSWER B: The white wire is the other side of your hot wire line cord. Don't touch it, either, even though we will consider it "neutral." It goes to one side of the transformer that kicks up the voltage inside your set. Remember, White for transformer Winding Without a fuse.

3AD-1-1.4 Why is the retaining screw in one terminal of a light socket made of brass while the other one is silver colored?
   A. To prevent galvanic action
   B. To indicate correct wiring polarity
   C. To better conduct current
   D. To reduce skin effect

ANSWER B: The silver thread is the neutral connection, and the brass button is the hot terminal inside your light socket. We use different metals to keep track of polarity.

3AD-1-2.1 How much electrical current flowing through the human body is usually fatal?
   A. As little as 100 milliamperes may be fatal.
   B. Approximately 10 amperes is required to be fatal.
   C. More than 20 amperes is needed to kill a human being.
   D. No amount of current will harm you. Voltages of over 2000 volts are
      always fatal, however.

ANSWER A: One-tenth of an ampere (amp) is the same as 100 milliamperes. One-tenth of an amp is as little as one tiny light on a dial lamp. This is still enough to zap you for good if it travels through your body and through your heart in an electrocution. *Never, never, never* work without shoes on a concrete garage floor. *Never* let any metal electrical appliance get near a bathtub.

3AD-1-2.2 What is the minimum voltage considered to be dangerous to humans?
   A. 30 volts
   B. 100 volts
   C. 1000 volts
   D. 2000 volts

ANSWER A: Even a couple of golf cart batteries could kill you if you aren't careful. This is why you must be especially careful when leaning across a bank of batteries and allowing current to flow through your body accidentally.

3AD-1-2.3 How much electrical current flowing through the human body is usually painful?
   A. As little as 50 milliamperes may be painful.
   B. Approximately 10 amperes is required to be painful.
   C. More than 20 amperes is needed to be painful to a human being.
   D. No amount of current will be painful. Voltages of over 2000 volts are
      always painful, however.

ANSWER A: We usually begin to feel the sensation of pain with as little as 50 milliamperes of current. This is less current than some electric clocks require. Be extremely careful never to let any current pass through your body. When it gets to the heart, it's all over for you.

3AD-1-3.1 Where should the main power-line switch for a high voltage power supply be situated?
A. Inside the cabinet, to interrupt power when the cabinet is opened
B. On the rear panel of the high-voltage supply
C. Where it can be seen and reached easily
D. This supply should not be switch-operated.
ANSWER C: Does everyone know where the main power switch is for your home? Let your whole family try turning off the power so they are familiar with the power line switch. If you have a special power switch in your ham shack, let everybody try their hand at turning it on and off.

3AD-2-1.1 How is a *voltmeter* typically connected to a circuit under test?
A. In series with the circuit
B. In parallel with the circuit
C. In quadrature with the circuit
D. In phase with the circuit
ANSWER B: We test for voltage by hooking our meter *across* the voltage source without undoing any wires. This is called a parallel connection. Checking the voltage when operating equipment is called "checking the voltage source under load."

3AD-2-2.1 How can the range of a *voltmeter* be extended?
A. By adding resistance in series with the circuit under test
B. By adding resistance in parallel with the circuit under test
C. By adding resistance in series with the meter
D. By adding resistance in parallel with the meter
ANSWER C: Let's say you want to test your new 12-volt power supply, but all you have is an old 6-volt panel meter. Through calculations (not required for this exam) you could add resistance in series with the meter to get 12 volts to register exactly 6 volts on your panel meter.

3AD-3-1.1 How is an *ammeter* typically connected to a circuit under test?
A. In series with the circuit
B. In parallel with the circuit
C. In quadrature with the circuit
D. In phase with the circuit
ANSWER A: An ammeter measures current. To measure current, disconnect the load to the direct battery connection and insert an ammeter *in series* with the circuit.

3AD-3-2.1 How can the range of an *ammeter* be extended?
A. By adding resistance in series with the circuit under test
B. By adding resistance in parallel with the circuit under test
C. By adding resistance in series with the meter
D. By adding resistance in parallel with the meter
ANSWER D: An ammeter measures the flow of electrons through its terminals. The keyword here is "through." We must remove a lead to the load we wish to test in an ammeter installation. However, if we wish to bypass some of that current to get our ammeter to read current well beyond its range, we would add resistance in parallel with the meter. This resistance, called a "shunt," allows some of the current to bypass the meter—thus giving the meter the capability to

read higher amounts of current without going off scale. Remember this—for the voltmeter, add resistance in series to extend it. For the ammeter, add resistance in parallel to extend it.

3AD-4.1 What is a multimeter?
A. An instrument capable of reading SWR and power
B. An instrument capable of reading resistance, capacitance and inductance
C. An instrument capable of reading resistance and reactance
D. An instrument capable of reading voltage, current and resistance

ANSWER D: Every amateur operator should own a multimeter. The multiple function meter can check everything from continuity to voltage, to current, to resistance. Even an inexpensive multimeter is better than no meter when you are trying to check out a circuit in the field. You can buy an excellent multimeter for less than $25.00 at your local Radio Shack store.

3AD-5-1.1 Where in the antenna transmission line should a peak-reading wattmeter be attached to determine the transmitter output power?
A. At the transmitter output
B. At the antenna feed point
C. One-half wavelength from the antenna feed point
D. One-quarter wavelength from the transmitter output

ANSWER A: The best place to test for power output is directly at the transmitter antenna terminal. Measuring it here eliminates any loss from the coax cable.

3AD-5-1.2 For the most accurate readings of transmitter output power, where should the rf wattmeter be inserted?
A. The wattmeter should be inserted and the output measured one-quarter wavelength from the antenna feed point.
B. The wattmeter should be inserted and the output measured one-half wavelength from the antenna feed point.
C. The wattmeter should be inserted and the output power measured at the transmitter antenna jack.
D. The wattmeter should be inserted and the output power measured at the Transmatch output.

ANSWER C: Here's another repeat question—and that is where the watt meter should go when testing power output on a transmitter. Put it on the transmitter. It's that simple.

3AD-5-1.3 At what line impedance are rf wattmeters usually designed to operate?
A. 25 ohms
B. 50 ohms
C. 100 ohms
D. 300 ohms

ANSWER B: Since Amateur Radio transceivers all operate at a 50-ohm impedance, our wattmeter should also be rated at 50 ohms. The impedance is kept the same for a perfect match.

3AD-5-1.4 What is a directional wattmeter?
A. An instrument that measures forward or reflected power
B. An instrument that measures the directional pattern of an antenna

C. An instrument that measures the energy consumed by the transmitter

D. An instrument that measures thermal heating in a load resistor

ANSWER A: A directional wattmeter is that special wattmeter that was mentioned earlier that can simultaneously measure forward and reflected power levels. Some directional wattmeters use only one needle, and a control that allows you to check for forward, and then reverse, power levels. They are a handy thing to have if you experiment with antennas. See figure at question 3AD-5-2.1.

3AD-5-2.1 If a directional rf wattmeter indicates 90 watts forward power and 10 watts reflected power, what is the actual transmitter output power?

A. 10 watts

B. 80 watts

C. 90 watts

D. 100 watts

ANSWER B: You can buy an inexpensive twin-needle watt meter that allows you to simultaneously monitor forward power and reverse power in watts. Any watts reflected back to the transmitter by a poor antenna are considered lost. They go up in transmitter heat.

Forward Power   −   Reflected Power   =   Actual Power Output
      90 W   −   10 W   =   80 W

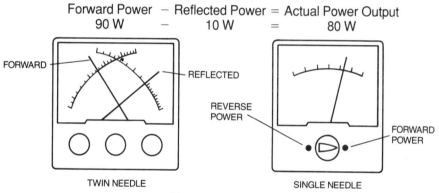

TWIN NEEDLE   SINGLE NEEDLE

Directional Watt Meter

3AD-5-2.2 If a directional rf wattmeter indicates 96 watts forward power and 4 watts reflected power, what is the actual transmitter output power?

A. 80 watts

B. 88 watts

C. 92 watts

D. 100 watts

ANSWER C: Simply subtract the wasted reflected power from the forward output power, and you can see what's actually reaching the antenna feedpoint.

Forward Power   −   Reflected Power   =   Actual Power Output
      96 W   −   4 W   =   92 W

3AD-6.1 What is a marker generator?

A. A high-stability oscillator that generates a series of reference signals at known frequency intervals

B. A low-stability oscillator that "sweeps" through a band of frequencies

C. An oscillator often used in aircraft to determine the craft's location relative to the inner and outer markers at airports

D. A high-stability oscillator whose output frequency and amplitude can be varied over a wide range

ANSWER A: Most worldwide transceivers have a built-in marker generator that sends out a tone, when turned on, at 25-kHz intervals. This can be useful in calibrating your equipment. It's also useful for the visually impaired. We use these tones as a reference signal.

3AD-6.2 What type of circuit is used to inject a frequency calibration signal into a communications receiver?
A. A product detector
B. A receiver incremental tuning circuit
C. A balanced modulator
D. A crystal calibrator

ANSWER D: On old tuning dial receivers, the dial cord would sometimes break. In order to find out exactly where we were, we would use a crystal calibrator to get us back on frequency. You won't find one of these in many modern sets today.

3AD-6.3 How is a marker generator used?
A. To calibrate the tuning dial on a receiver
B. To calibrate the volume control on a receiver
C. To test the amplitude linearity of an SSB transmitter
D. To test the frequency deviation of an FM transmitter

ANSWER A: Here's that marker generator again—this time they want to know how it is used. Use it to mark the band edges when operating mobile. If you use it correctly, it allows you to calibrate the tuning dial all in your head without having to look.

3AD-7.1 What piece of test equipment produces a stable, low-level signal that can be set to a specific frequency?
A. A wavemeter
B. A reflectometer
C. A signal generator
D. A balanced modulator

ANSWER C: A signal generator with a digital readout is useful in troubleshooting scanner radios. You can inject a tiny signal into the "front end" of the set to check it's operation and to be sure it is on frequency.

3AD-7.2 What is an RF signal generator commonly used for?
A. Measuring RF signal amplitude
B. Aligning receiver tuned circuits
C. Adjusting the transmitter impedance-matching network
D. Measuring transmission line impedance

ANSWER B: Your author uses an RF signal generator in his shack for aligning scanners, shortwave sets, and hand-held portables.

3AD-8-1.1 What is a reflectometer?
A. An instrument used to measure signals reflected from the ionosphere
B. An instrument used to measure radiation resistance
C. An instrument used to measure transmission-line impedance
D. An instrument used to measure standing wave ratio

ANSWER D: When testing antennas, use a meter that measures reflected power. The reflectometer is simply another instrument that lets you measure the standing wave ratio (SWR)—or the ratio of forward power to reflected power. You want the SWR as low as possible.

3AD-8-1.2 What is the device that can indicate an impedance mismatch in an antenna system?
A. A field-strength meter
B. A set of lecher wires
C. A wavemeter
D. A reflectometer

ANSWER D: Impedance is a word that describes the opposition to the flow of electrons in a circuit. Most circuits have a specific impedance, and most Amateur Radio antennas should have an impedance of 50 ohms. If your antenna is perfectly matched to your coax and transceiver, everything will be a 50-ohm impedance match, and you will have minimum reflected power. You can verify this with an SWR reflectometer.

3AD-8-2.1 For best accuracy when adjusting the impedance match between an antenna and feed line, where should the match-indicating device be inserted?
A. At the antenna feed point
B. At the transmitter
C. At the midpoint of the feed line
D. Anywhere along the feed line

ANSWER A: If you do a lot of experimenting with antennas, and we hope you will, buy an external SWR meter. Although most worldwide radios have a built-in SWR meter, it measures the SWR after it travels all the way back down the feed line. What you want is a separate SWR meter so you can take it to the antenna feed point, and measure SWR truly presented by the antenna itself.

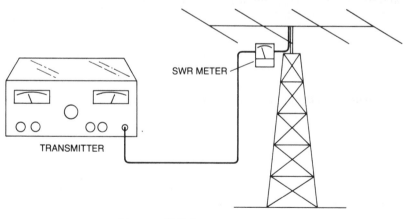

SWR METER

TRANSMITTER

Measure SWR Right at Antenna

3AD-8-2.2 Where should a reflectometer be inserted into a long antenna transmission line in order to obtain the most valid standing wave ratio indication?
A. At any quarter-wavelength interval along the transmission line
B. At the receiver end
C. At the antenna end
D. At any even half-wavelength interval along the transmission line

ANSWER C: Once again, if you do your own antenna work, buy an external SWR meter so you can measure what's happening right at the antenna.

3AD-9.1 When adjusting a transmitter filter circuit, what device is connected to the transmitter output?
  A. A multimeter
  B. A set of Litz wires
  C. A receiver
  D. A dummy antenna
ANSWER D: And here's that dummy antenna question again—it's used when you want to transmit full power output, but not have it go beyond a block or so away.

3AD-9.2 What is a dummy antenna?
  A. An isotropic radiator
  B. A nonradiating load for a transmitter
  C. An antenna used as a reference for gain measurements
  D. The image of an antenna, located below ground
ANSWER B: Here's that dummy antenna question, again! I guess they really want you to know it for the examination. A dummy antenna does not radiate very far. However, you might be surprised how far you can talk on a lightbulb. Your author, Gordon West, once communicated 3000 miles away on a 100-watt lightbulb, much to the amazement of him and his ham radio students.

3AD-9.3 Of what materials may a dummy antenna be made?
  A. A wire-wound resistor
  B. A diode and resistor combination
  C. A noninductive resistor
  D. A coil and capacitor combination
ANSWER C: You can buy a dummy load that will handle up to 100 watts for about $20. You could also make it up yourself if you found some big resistors that don't use a coil of wire as their resistor windings. I like lightbulbs because I can see immediately the results—however, lightbulbs tend to act like an inductor (a coil of wire), so they are not the best. Any non-inductive resistor combination will do the job nicely.

3AD-9.4 What station accessory is used in place of an antenna during transmitter tests so that no signal is radiated?
  A. A Transmatch
  B. A dummy antenna
  C. A low-pass filter
  D. A decoupling resistor
ANSWER B: Here's yet another question about that antenna that doesn't radiate— the dummy load or dummy antenna.

3AD-9.5 What is the purpose of a dummy load?
  A. To allow off-the-air transmitter testing
  B. To reduce output power for QRP operation
  C. To give comparative signal reports
  D. To allow Transmatch tuning without causing interference
ANSWER A: Another question about the dummy load—just remember, it doesn't radiate, and allows you to run your transmitter full bore without causing interference.

3AD-9.6 How many watts should a dummy load for use with a 100-watt single-sideband phone transmitter be able to dissipate?
A. A minimum of 100 watts continuous
B. A minimum of 141 watts continuous
C. A minimum of 175 watts continuous
D. A minimum of 200 watts continuous
ANSWER A: Your single-sideband phone transmitter will put out about 100 watts. This means you need a dummy load capable of handling that power—100 watts continuous.

3AD-10.1 What is an S-meter?
A. A meter used to measure sideband suppression
B. A meter used to measure spurious emissions from a transmitter
C. A meter used to measure relative signal strength in a receiver
D. A meter used to measure solar flux
ANSWER C: Remember the S in RST? That stands for signal strength, and we use an S meter to measure relative signal strength in a communications receiver. Almost all types of Amateur Radio sets have some sort of an S meter. Even the handhelds have a relative signal strength bar-graph display.

3AD-10.2 A meter that is used to measure relative signal strength in a receiver is known as what?
A. An S-meter
B. An RST-meter
C. A signal deviation meter
D. An SSB meter
ANSWER A: An S-meter measures signal strength. All worldwide ham sets have one.

3AD-11-1.1 Large amounts of RF energy may cause damage to body tissue, depending on the wavelength of the signal, the energy density of the RF field, and other factors. How does RF energy effect body tissue?
A. It causes radiation poisioning.
B. It heats the tissue.
C. It cools the tissue.
D. It produces genetic changes in the tissue.
ANSWER B: It's the heating effect on body tissue that causes permanent damage.

3AD-11-1.2 Which body organ is most susceptible to damage from the heating effects of radio frequency radiation?
A. Eyes
B. Hands
C. Heart
D. Liver
ANSWER A: The eyes are most sensitive to RF radiation.

3AD-11-2.1 Scientists have devoted a great deal of effort to determine safe RF exposure limits. What organization has established an RF protection guide?
A. The Institute of Electrical and Electronics Engineers
B. The American Radio Relay League
C. The Environmental Protection Agency
D. The American National Standards Institute

ANSWER D: This organization is the authority on RF protection. You will see their acronym, ANSI, often on standards and specifications.

3AD-11-2.2 What is the purpose of the ANSI RF protection guide?
A. It protects you from unscrupulous radio dealers.
B. It sets RF exposure limits under certain circumstances.
C. It sets transmitter power limits.
D. It sets antenna height requirements.
ANSWER B: Watch out when you are working with antennas. RF protection is essential.

3AD-11-2.3 The American National Standards Institute RF protection guide sets RF exposure limits under certain circumstances. In what frequency range is the maximum exposure level the most stringent (lowest)?
A. 3 to 30 MHz
B. 30 to 300 MHz
C. 300 to 3000 MHz
D. Above 1.5 GHz
ANSWER B: When you get up to the VHF bands, exposure to RF gets to be a big issue.

3AD-11-2.4 The American National Standards Institute RF protection guide sets RF exposure limits under certain circumstances. Why is the maximum exposure level the most stringent (lowest) in the ranges between 30 MHz and 300 MHz?
A. There are fewer transmitters operating in this frequency range.
B. There are more transmitters operating in this frequency range.
C. Most transmissions in this frequency range are for an extended time.
D. Human body lengths are close to whole-body resonance in that range.
ANSWER D: Within our body, we have many small organs and muscles with a length that corresponds to VHF wavelength frequencies. This is why we must be concerned with RF exposure between 30 MHz and 300 MHz—our body parts are actually resonant!

3AD-11-2.5 The American National Standards Institute RF protection guide sets RF exposure limits under certain circumstances. What is the maximum safe power output to the antenna terminal of a hand-held VHF or UHF radio, as set by this RF protection guide?
A. 125 milliwatts
B. 7 watts
C. 10 watts
D. 25 watts
ANSWER B: Most VHF and UHF handhelds don't put out much over 5 watts, which is 2 watts under the 7-watt limit.

3AD-11-3.1 After you make internal tuning adjustments to your VHF power amplifier, what should you do before you turn the amplifier on?
A. Remove all amplifier shielding to ensure maximum cooling.
B. Connect a noise bridge to eliminate any interference.
C. Be certain all amplifier shielding is fastened in place.
D. Be certain no antenna is attached so that you will not cause any interference.

ANSWER C: Transmitter shielding is important to prevent unwanted radiation of RF signals to your body.

Subelement 3AE – Electrical Principles (2 examination questions from 34 questions in 3AE)

3AE-1-1.1 What is meant by the term *resistance?*
A. The opposition to the flow of current in an electric circuit containing inductance
B. The opposition to the flow of current in an electric circuit containing capacitance
C. The opposition to the flow of current in an electric circuit containing reactance
D. The opposition to the flow of current in an electric circuit that does not contain reactance

ANSWER D: Think of resistance as stepping on a garden hose. This restricts the water flow, similar to a resistor that restricts the flow of electrons in a direct current (dc) circuit.

R

a. Symbol

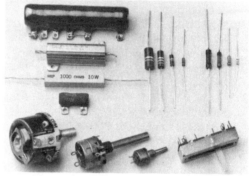

b. Physical Part

Resistor

3AE-1-2.1 What is an ohm?
A. The basic unit of resistance
B. The basic unit of capacitance
C. The basic unit of inductance
D. The basic unit of admittance

ANSWER A: We quantify resistance by the unit "ohm." A circuit component has a resistance of one ohm when one volt applied to the component produces a current of one ampere. There are special types of impedances in alternating current circuits that act like resistances in direct current circuits. They are called inductive and capacitive reactance, and are also quantified in the ohm as a unit of impedance.

3AE-1-2.2 What is the unit measurement of resistance?
A. Volt
B. Ampere
C. Joule
D. Ohm

ANSWER D: When you think of resistance, remember the ohm.

3AE-1-3.1 Two equal-value resistors are connected *in series*. How does the total resistance of this combination compare with the value of either resistor by itself?
  A. The total resistance is half the value of either resistor.
  B. The total resistance is twice the value of either resistor.
  C. The total resistance is the same as the value of either resistor.
  D. The total resistance is the square of the value of either resistor.
ANSWER B: Resistors *in series* simply add up. If they are equal, their values will be twice the value of a single resistor.

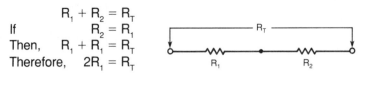

If $\quad\quad R_1 + R_2 = R_T$
$\quad\quad\quad\quad R_2 = R_1$
Then, $\quad R_1 + R_1 = R_T$
Therefore, $\quad 2R_1 = R_T$

Makes Larger Resistance

3AE-1-3.2 How does the total resistance of a string of series-connected resistors compare to the values of the individual resistors?
  A. The total resistance is the square of the sum of all the individual resistor values.
  B. The total resistance is the square root of the sum of the individual resistor values.
  C. The total resistance is the sum of the squares of the individual resistor values.
  D. The total resistance is the sum of all the individual resistance values.
ANSWER D: Resistors *in series* simply add up.

$$R_1 + R_2 + R_3 + \ldots + R_N = R_T$$

3AE-1-4.1 Two equal-value resistors are connected *in parallel*. How does the total resistance of this combination compare with the value of either resistor by itself?
  A. The total resistance is twice the value of either resistor.
  B. The total resistance is half the value of either resistor.
  C. The total resistance is the square of the value of either resistor.
  D. The total resistance is the same as the value of either resistor.
ANSWER B: When two identical value resistors are *in parallel,* dividing the value of one of the resistors by two gives the total resistance. The general equation for finding the equivalent resistance of *any* two resistors in parallel is:

$$\frac{R_1 \times R_2}{R_1 + R_2} = R_T$$

If $\quad\quad R_2 = R_1 \quad$ Then $\dfrac{R_1 \times R_1}{R_1 + R_1} = R_T$

Therefore, $\dfrac{R_1^2}{2R_1} = R_T \quad$ And $\quad \dfrac{R_1}{2} = R_T$

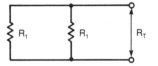

Makes Smaller Resistance

3AE-1-4.2 How does the total resistance of a string of parallel-connected resistors compare to the values of the individual resistors?
A. The total resistance is the square of the sum of the resistor values.
B. The total resistance is more than the highest-value resistor in the combination.
C. The total resistance is less than the smallest-value resistor in the combination.
D. The total resistance is same as the highest-value resistor in the combination.

ANSWER C: This is a quick and easy way to estimate the total resistance of a string of parallel connected resistors—it will always be less than the value of the lowest value resistor.

$$\frac{1}{R_1} + \frac{1}{R_2} + \frac{1}{R_3} + ... + \frac{1}{R_n} = \frac{1}{R_T}$$

3AE-2.1 What is Ohm's Law?
A. A mathematical relationship between resistance, voltage and power in a circuit
B. A mathematical relationship between current, resistance and power in a circuit
C. A mathematical relationship between current, voltage and power in a circuit
D. A mathematical relationship between resistance, current and applied voltage in a circuit

ANSWER D: Ohm's Law describes the relationship between voltage (E), current (I), and resistance (R). If you draw this magic circle, it's easy to calculate any question with Ohms Law:

For Voltage: $E = I \times R$
For Current: $I = \frac{E}{R}$
For Resistance: $R = \frac{E}{I}$

E = VOLTAGE IN VOLTS
I = CURRENT IN AMPERES
R = RESISTANCE IN OHMS

Ohm's Law Calculation

In the circle we have E over IR. Whatever you are trying to solve for, remove it from the circle. You can do this effectively by putting your finger over the letter for which you are solving. Now plug in the values that they give you on the Technician test. They will be exactly the same as the following questions in this book. Solve the problem working the remaining letters as they remain in the circle.

3AE-2.2 How is the current in a dc circuit calculated when the voltage and resistance are known?
A. $I = E / R$
B. $P = I \times E$
C. $I = R \times E$
D. $I = E \times R$

ANSWER A: Use the magic circle from question 3AE-2.1. In this question, I (current) is equal to E (voltage) divided by R (resistance).

3AE-2.3 What is the input resistance of a load when a 12-volt battery supplies 0.25 amperes to it?
 A. 0.02 ohms
 B. 3 ohms
 C. 48 ohms
 D. 480 ohms

ANSWER C: This question asks for the resistance. Pull R out of the circle, and substitute the given numbers for the letters. Substitute 12 for E and 0.25 for I. Since R is equal to E over I, R is now equal to 12 divided by 0.25 or R = 12/.25 You may do this longhand or you may use a calculator on the Technician test. Your answer should come out 48 ohms. Here are the calculator keystrokes: Clear 12 ÷ .25 = and the answer is 48.

3AE-2.4 The product of the current and what force gives the electrical power in a circuit?
 A. Magnetomotive force
 B. Centripetal force
 C. Electrochemical force
 D. Electromotive force

ANSWER D: Electromotive force is voltage. In Ohm's Law, E always stands for voltage. Watch out for answer C—it looks close, but it is not the right answer. Another magic circle that allows us to calculate power if current and voltage are known shows that P = E × I. Or if P is known and I is known, E = P/I. Likewise, I = P/E, if P and E are known. Use this circle the same as the Ohm's Law circle of question 3AE-2.1.

For Power: $P = E \times I$
For Voltage: $E = \dfrac{P}{I}$
For Current: $I = \dfrac{P}{E}$

P = POWER IN WATTS
E = VOLTAGE IN VOLTS
I = CURRENT IN AMPERES

Power Calculation

3AE-2.5 What is the input resistance of a load when a 12-volt battery supplies 0.15 amperes to it?
 A. 8 ohms
 B. 80 ohms
 C. 100 ohms
 D. 800 ohms

ANSWER B: Refer to question 3AE-2.1. Write down that Ohm's Law circle! Learn it well and be able to apply it easily. This problem is similar to question 3AE-2.3, but in this case the current is only 0.15 ampere. As a result, R is equal to 12 divided by 0.15, or R = 12/.15. Your answer is 80 ohms. The calculator keystrokes are: Clear 12 ÷ .15 = and the answer is 80.

3AE-2.6 When 120 volts is measured across a 4700-ohm resistor, approximately how much current is flowing through it?
 A. 39 amperes
 B. 3.9 amperes

C. 0.26 ampere

D. 0.026 ampere

ANSWER D: Since they want to know how much current is present, remove I from the Ohm's Law magic circle of question 3AE-2.1. When I is covered, you have E/R remaining. You must divide E by R, so divide 4700 into 120. As you can see, it doesn't divide evenly. Don't reverse it, or you'll end up with the wrong answer, and as a temptation, the wrong answer is one of the choices! They are just waiting for you to do it backwards. Use your calculator and divide 4700 into 120, and you should end up with 0.026 amp. Here are the calculator keystrokes: Clear 120 ÷ 4700 = and the answer is .0255. Round it to .026 ampere.

3AE-2.7 When 120 volts is measured across a 47000-ohm resistor, approximately how much current is flowing through it?

A. 392 A

B. 39.2 A

C. 26 mA

D. 2.6 mA

ANSWER D: Here we go again. Use the same magic circle E/R because you want to find current, but the resistance value has an additional zero. You now divide 120 by 47000. You get 0.00255 ampere. However, you must change amps to milliamps (10^{-3}) in order to arrive at a listed answer. Move the decimal point three places to the right, and there it is—2.6 mA (rounded off). However, if you do the problem backwards, you'll find that they also have an incorrect answer waiting for you. Watch out! Here are the calculator keystrokes: Clear 120 ÷ 47000 =. Again, when the decimal point is moved three places to the right, the rounded answer is 2.6 mA.

3AE-2.8 When 12 volts is measured across a 4700-ohm resistor, approximately how much current is flowing through it?

A. 2.6 mA

B. 26 mA

C. 39.2 A

D. 392 A

ANSWER A: And here we go again, but with some different zeros in the question—R is 4700 and E is 12. Use your calculator to divide 4700 into 12. Don't do it backwards! Now take your answer and move the decimal point three places to the right to change amps to milliamps. When you express your answer in milliamperes, it means you must multiply the answer by 10^{-3} to arrive at the number of amperes. Here are the calculator keystrokes: Clear 12 ÷ 4700 =. Again move the decimal point and round the .00255 to 2.6 mA.

3AE-2.9 When 12 volts is measured across a 47000-ohm resistor, approximately how much current is flowing through it?

A. 255 µA

B. 255 mA

C. 3917 mA

D. 3917 A

ANSWER A: Same numbers, but again additional zeros. Use your calculator to divide 47000 into 12. 12 divided by 47000 is 0.000255. The correct answer will be found in microamps (10^{-6}). This requires you to move the decimal point six places to the right. Here are the calculator keystrokes: Clear 12 ÷ 47000 =. You end up with 255 µA when the decimal point is moved six places to the right.

3AE-3-1.1 What is the term used to describe the ability of a component to store energy in a magnetic field?
A. Admittance
B. Capacitance
C. Inductance
D. Resistance
ANSWER C: Place a magnetic compass near an energized inductor and watch what happens! The compass needle lines up with the magnetic field, which is the basis for inductance.

3AE-3-2.1 What is the basic unit of inductance?
A. Coulomb
B. Farad
C. Henry
D. Ohm
ANSWER C: It is the henry, but because the henry is a fairly large unit, we measure inductance in one thousandths of a henry (millihenry) and one millionths of a henry (microhenry).

3AE-3-2.2 What is a henry?
A. The basic unit of admittance
B. The basic unit of capacitance
C. The basic unit of inductance
D. The basic unit of resistance
ANSWER C: When you think of inductance, think of the henry.

3AE-3-2.3 What is a microhenry?
A. A basic unit of inductance equal to 10^{-12} henrys
B. A basic unit of inductance equal to 10^{-6} henrys
C. A basic unit of inductance equal to 10^{-3} henrys
D. A basic unit of inductance equal to 10^{6} henrys
ANSWER B: Now let's get back to those coils. Inductance is measured in a unit called the henry. A single henry is usually too large, so we break it down to the microhenry, or millihenry. A microhenry is one millionth of a henry and is noted 10^{-6} henry.

Prefix	Symbol		Multiplication Factor	
giga	G	1×10^{9}	1,000,000,000	one billion
mega	M	1×10^{6}	1,000,000	one million
kilo	k	1×10^{3}	1,000	one thousand
centi	c	1×10^{-2}	0.01	one hundredth
milli	m	1×10^{-3}	0.001	one thousandth
micro	μ	1×10^{-6}	0.000001	one millionth
nano	n	1×10^{-9}	0.000000001	one billionth
pico	p	1×10^{-12}	0.000000000001	one trillionth

International Unit of Measurement Prefixes

3AE-3-2.4 What is a millihenry?
A. A basic unit of inductance equal to 10^{-12} henrys.
B. A basic unit of inductance equal to 10^{-6} henrys.
C. A basic unit of inductance equal to 10^{-3} henrys.
D. A basic unit of inductance equal to 10^{6} henrys.
ANSWER C: A millihenry is one thousandth of a henry, and is noted 10^{-3}.

3AE-3-3.1 Two equal-value inductors are connected *in series*. How does the total inductance of this combination compare with the value of either inductor by itself?
  A. The total inductance is half the value of either inductor.
  B. The total inductance is twice the value of either inductor.
  C. The total inductance is equal to the value of either inductor.
  D. No comparison can be made without knowing the exact inductances.
ANSWER B: Inductors are treated like resistors with respect to total value calculations. Their values add *in series*.

$$L_1 + L_2 = L_T$$

3AE-3-3.2 How does the total inductance of a string of series-connected inductors compare to the values of the individual inductors?
  A. The total inductance is equal to the average of all the individual inductances.
  B. The total inductance is equal to less than the value of the smallest inductance.
  C. The total inductance is equal to the sum of all the individual inductances.
  D. No comparison can be made without knowing the exact inductances.
ANSWER C: Just pretend that the inductor problem is a resistor problem. They simply add up.

$$L_1 + L_2 + L_3 + ... + L_N = L_T$$

3AE-3-4.1 Two equal-value inductors are connected *in parallel*. How does the total inductance of this combination compare with the value of either inductor by itself?
  A. The total inductance is half the value of either inductor.
  B. The total inductance is twice the value of either inductor.
  C. The total inductance is equal to the square of either inductance.
  D. No comparison can be made without knowing the exact inductances.
ANSWER A: Inductors *in parallel* are treated just like resistors in parallel with respect to total value calculations. When $L_1 = L_2$, $L_T = L_2/2$.

$$\frac{L_1 \times L_2}{L_1 + L_2} = L_T$$

3AE-3-4.2 How does the total inductance of a string of parallel-connected inductors compare to the values of the individual inductors?
  A. The total inductance is equal to the sum of the inductances in the combination.
  B. The total inductance is less than the smallest inductance value in the combination.
  C. The total inductance is equal to the average of the inductances in the combination.
  D. No comparison can be made without knowing the exact inductances.
ANSWER B: See? Just like resistors! L_T will always be less than smallest L.

$$\frac{1}{L_1} + \frac{1}{L_2} + \frac{1}{L_3} + ... + \frac{1}{L_n} = \frac{1}{L_T}$$

3AE-4-1.1 What is the term used to describe the ability of a component to store energy in an electric field?
A. Capacitance
B. Inductance
C. Resistance
D. Tolerance

ANSWER A: Capacitors store their energy in an electric field, not a magnetic field like inductors. This is the basis for capacitance.

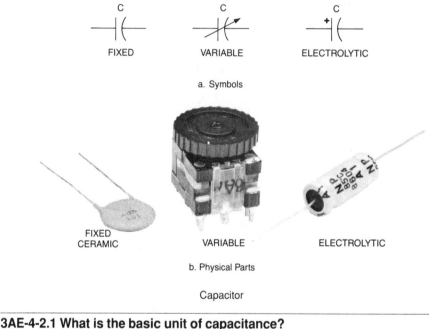

a. Symbols

b. Physical Parts

Capacitor

3AE-4-2.1 What is the basic unit of capacitance?
A. Farad
B. Ohm
C. Volt
D. Ampere

ANSWER A: Farad for capacitance, and ohm for resistance.

3AE-4-2.2 What is a microfarad?
A. A basic unit of capacitance equal to 10^{-12} farads
B. A basic unit of capacitance equal to 10^{-6} farads
C. A basic unit of capacitance equal to 10^{-2} farads
D. A basic unit of capacitance equal to 10^{6} farads

ANSWER B: The unit of measurement for capacitance is a farad. Normally capacitance is used in much smaller units than the farad, so become familiar with the word "microfarad." This is one millionth of a farad. It is noted 10^{-6}. In scientific notation, this 10^{-6} means 0.000001 farad.

3AE-4-2.3 What is a picofarad?
A. A basic unit of capacitance equal to 10^{-12} farads
B. A basic unit of capacitance equal to 10^{-6} farads

C. A basic unit of capacitance equal to 10^{-2} farads

D. A basic unit of capacitance equal to 10^6 farads

ANSWER A: A micromicrofarad is called a picofarad. It is noted 10^{-12}, and is a millionth of a millionth of a farad. Oldtimers sometimes refer to this as a "mickey mike." A picofarad equals 0.000000000001 farad.

3AE-4-2.4 What is a farad?

A. The basic unit of resistance

B. The basic unit of capacitance

C. The basic unit of inductance

D The basic unit of admittance

ANSWER B: When you think of capacitance, remember the farad.

3AE-4-3.1 Two equal-value capacitors are connected *in series.* How does the total capacitance of this combination compare with the value of either capacitor by itself?

A. The total capacitance is twice the value of either capacitor.

B. The total capacitance is equal to the value of either capacitor.

C. The total capacitance is half the value of either capacitor.

D. No comparison can be made without knowing the exact capacitances.

ANSWER C: Capacitor values combine just opposite of resistors. The general equation for finding the equivalent capacitance of any two capacitors *in series* is:

$$\frac{C_1 \times C_2}{C_1 + C_2} = C_T$$

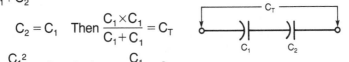

If $\quad C_2 = C_1 \quad$ Then $\dfrac{C_1 \times C_1}{C_1 + C_1} = C_T$

Therefore, $\dfrac{C_1{}^2}{2C_1} = C_T \quad$ And $\quad \dfrac{C_1}{2} = C_T \qquad$ Makes Smaller Capacitor

3AE-4-3.2 How does the total capacitance of a string of series-connected capacitors compare to the values of the individual capacitors?

A. The total capacitance is equal to the sum of the capacitances in the combination.

B. The total capacitance is less than the smallest value of capacitance in the combination.

C. The total capacitance is equal to the average of the capacitances in the combination.

D. No comparison can be made without knowing the exact capacitances.

ANSWER B: Combining capacitors in series is like combining resistors in parallel.

$$\frac{1}{C_1} + \frac{1}{C_2} + \frac{1}{C_3} + \dots + \frac{1}{C_n} = \frac{1}{C_T}$$

3AE-4-4.1 Two equal-value capacitors are connected *in parallel*. How does the total capacitance of this combination compare with the value of either capacitor by itself?
   A. The total capacitance is twice the value of either capacitor.
   B. The total capacitance is half the value of either capacitor.
   C. The total capacitance is equal to the value of either capacitor.
   D. No comparison can be made without knowing the exact capacitances.
ANSWER A: Treat capacitors just the opposite of what you would do with resistors. Capacitors *in parallel* add up to make larger capacitors.

$$C_1 + C_2 = C_T$$
If $$C_2 = C_1$$
Then, $$C_1 + C_1 = C_T$$
Therefore, $$2C_1 = C_T$$

Makes Larger Capacitor

3AE-4-4.2 How does the total capacitance of a string of parallel-connected capacitors compare to the values of the individual capacitors?
   A. The total capacitance is equal to the sum of the capacitances in the combination.
   B. The total capacitance is less than the smallest value of capacitance in the combination.
   C. The total capacitance is equal to the average of the capacitances in the combination.
   D. No comparison can be made without knowing the exact capacitances.
ANSWER A: Capacitors in parallel are treated like resistors in series—the values add up.

$$C_1 + C_2 + C_3 + \ldots + C_N = C_T$$

Subelement 3AF – Circuit Components (2 examination questions from 37 questions in 3AF)

3AF-1-1.1 What are the four common types of resistor construction?
   A. Carbon-film, metal-film, micro-film and wire-film
   B. Carbon-composition, carbon-film, metal-film and wire-wound
   C. Carbon-composition, carbon-film, electrolytic and metal-film
   D. Carbon-film, ferrite, carbon-composition and metal-film
ANSWER B: File these resistor types in your memory.

3AF-1-2.1 What is the primary function of a resistor?
   A. To store an electric charge
   B. To store a magnetic field
   C. To match a high-impedance source to a low-impedance load
   D. To limit the current in an electric circuit
ANSWER D: Resistors limit the current in a circuit. Remember this question from your Novice test?

3AF-1-2.2 What is a variable resistor?

A. A resistor that changes value when an ac voltage is applied to it
B. A device that can transform a variable voltage into a constant voltage
C. A resistor with a slide or contact that makes the resistance adjustable
D. A resistor that changes value when it is heated

ANSWER C: The volume control on your worldwide radio is a variable resistor.

a. Symbol

b. Physical Part

Variable Resistor

3AF-1-3.1 What do the first three color bands on a resistor indicate?

A. The value of the resistor in ohms
B. The resistance tolerance in percent
C. The power rating in watts
D. The value of the resistor in henrys

ANSWER A: For the Technician Class examination, you do not need to memorize the resistor color code. However, to be a good Amateur Radio experimenter, it's handy to have it memorized. See figure at question 3AF-1-3.2.

3AF-1-3.2 How can a carbon resistor's electrical tolerance rating be found?

A. By using a wavemeter
B. By using the resistor's color code
C. By using Thevenin's theorem for resistors
D. By using the Baudot code

ANSWER B: It's easy to read a resistor's color code. The first three bands tell us the resistor's resistance. The fourth band gives us the resistor's tolerance. ·

Here is how you read the first three bands—the first color tells us the first number, the second color the second number, and the third color tells us how many zeros to add after those first two numbers. Black is zero, brown is one, red is two, orange is three, yellow is four, green is five, blue is six, violet is seven, gray is eight, and white is nine. Red/black/red would then equal 2,000 ohms.

If the fourth band is gold, the tolerance is good at ±5 percent. Silver is an okay ±10 percent tolerance; and if there is no fourth band, the tolerance is assumed to be a ±20 percent average resistor.

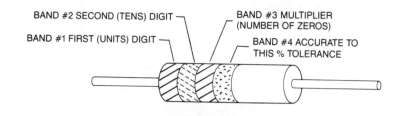

BAND #2 SECOND (TENS) DIGIT

BAND #1 FIRST (UNITS) DIGIT

BAND #3 MULTIPLIER
(NUMBER OF ZEROS)

BAND #4 ACCURATE TO
THIS % TOLERANCE

EXAMPLES

Band	1	2	3	4	Value
Resistor #1	Red (2)	Black (0)	Red (00)	Gold 5%	2,000 ±5%
Resistor #2	Brown (1)	Black (0)	Green (00000)	—	1,000,000 ±20%
Resistor #3	Blue (6)	White (9)	Orange (000)	Silver 10%	69,000 ±10%

Resistor Color Code (First Three Bands)   **Tolerance** (Fourth Band)

Black	0	Blue	6		Gold	±5%
Brown	1	Violet	7		Silver	±10%
Red	2	Gray	8		None	±20%
Orange	3	White	9			
Yellow	4	Silver	0.01			
Green	5	Gold	0.1			

Resistor Color Code

Source: Technology Dictionary ©1987 Master Publishing, Inc., Richardson, Texas

3AF-1-3.3 What does the fourth color band on a resistor indicate?

A. The value of the resistor in ohms
B. The resistance tolerance in percent
C. The power rating in watts
D. The resistor composition

ANSWER B: Remember that gold is 5 percent tolerance, and that is one of the best. See figure at question 3AF-1-3.2.

3AF-1-3.4 When the color bands on a group of resistors indicate that they all have the same resistance, what further information about each resistor is needed in order to select those that have nearly equal value?

A. The working voltage rating of each resistor
B. The composition of each resistor
C. The tolerance of each resistor
D. The current rating of each resistor

ANSWER C: It's possible to purchase resistors with tolerances as good as 5 percent. A molded carbon resitor has a gold fourth band to indicate 5 percent tolerance. For some resistors, the tolerance is printed numerically on the resistor's body.

3AF-1-4.1 Why do resistors generate heat?

A. They convert electrical energy to heat energy.
B. They exhibit reactance.
C. Because of skin effect
D. To produce thermionic emission

ANSWER A: Resistors get hot as they burn up energy.

3AF-1-4.2 Why would a large size resistor be substituted for a smaller one of the same resistance?

    A. To obtain better response
    B. To obtain a higher current gain
    C. To increase power dissipation capability
    D. To produce a greater parallel impedance

ANSWER C: Besides the actual resistance and tolerance of a resistor, its power dissipation plays an important part in resistor selection. Larger resistors have greater power dissipation capabilities.

3AF-1-5.1 What is the symbol used to represent a fixed resistor on schematic diagrams?

A.   ⌒⌒⌒   B.   ⌇⌇⌇   C.   —)⊢   D.   —▶◀—

ANSWER B: Resistance to current is similar to a rough surface that presents friction to another surface rubbing against it. The squiggly line symbol should remind you of a rough surface.

3AF-1-5.2 What is the symbol used to represent a variable resistor on schematic diagrams

A.   ⤢⌒   B.   ⊁⊢   C.   ⤨⌇   D.   ⌇↑

ANSWER C: Same resistor symbol, but with an arrow through it.

3AF-2-1.1 What is an inductor core?

    A. The point at which an inductor is tapped to produce resonance
    B. A tight coil of wire used in a transformer
    C. An insulating material placed between the plates of an inductor
    D. The central portion of a coil; may be made from air, iron, brass or other material

ANSWER D: Iron cores are very fragile—if you must adjust them, use the proper tool, and adjust them carefully. They are brittle.

3AF-2-1.2 What are the component parts of a coil?

    A. The wire in the winding and the core material
    B. Two conductive plates and an insulating material
    C. Two or more layers of silicon material
    D. A donut-shaped iron core and a layer of insulating tape

ANSWER A: Wire in the windings and the coil core are the two main parts of a coil.

3AF-2-1.3 Describe an inductor.

    A. A semiconductor in a conducting shield
    B. Two parallel conducting plates
    C. A straight wire conductor mounted inside a Faraday shield
    D. A coil of conducting wire

ANSWER D: When you see the word "inductor," always think of a coil of wire. Anytime a piece of wire is made into a coil in a circuit, inductance is being added to the circuit.

L

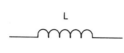

a. Symbol

b. Physical Part

Inductor

3AF-2-1.4 For radio frequency power applications, which type of inductor has the least amount of loss?
A. Magnetic wire
B. Iron core
C. Air core
D. Slug tuned

ANSWER C: In all ham radio transmitter sections, you will usually find open wound coils with nothing in their core—well, nothing but air. Be careful not to distort the shape of these coils because they are wound for the precise right inductance to make the final amplifier circuit work properly.

3AF-2-2.1 What is an inductor?
A. An electronic component that stores energy in an electric field
B. An electronic component that converts a high voltage to a lower voltage
C. An electronic component that opposes dc while allowing ac to pass
D. An electronic component that stores energy in a magnetic field

ANSWER D: Remember that coils develop a magnetic field which is indicated by a magnetic compass held near the energized coil.

3AF-2-2.2 What are the electrical properties of an inductor?
A. An inductor stores a charge electrostatically and opposes a change in voltage.
B. An inductor stores a charge electrochemically and opposes a change in current.
C. An inductor stores a charge electromagnetically and opposes a change in current.
D. An inductor stores a charge electromechanically and opposes a change in voltage.

ANSWER C: The key term in the answer is "magnetically."

3AF-2-3.1 What factors determine the amount of inductance in a coil?
A. The type of material used in the core, the diameter of the core and whether the coil is mounted horizontally or vertically
B. The diameter of the core, the number of turns of wire used to wind the coil and the type of metal used in the wire
C. The type of material used in the core, the number of turns used to wind the core and the frequency of the current through the coil
D. The type of material used in the core, the diameter of the core, the length of the coil and the number of turns of wire used to wind the coil

ANSWER D: All these factors determine the inductance of a coil. Iron within the coil adds inductance; brass reduces inductance.

3AF-2-3.2 What can be done to raise the inductance of a 5-microhenry air-core coil to a 5-millihenry coil with the same physical dimensions?
   A. The coil can be wound on a non-conducting tube.
   B. The coil can be wound on an iron core.
   C. Both ends of the coil can be brought around to form the shape of a donut, or toroid.
   D. The coil can be made of a heavier-gauge wire.
ANSWER B: Since a 5-millihenry coil has more inductance than a 5-microhenry coil, we can create this greater inductance by inserting an iron core. Most iron cores are adjustable. By moving the core, the coil inductance can be adjusted to the proper value. *Iron* in spinach makes you *stronger!*

3AF-2-3.3 As an iron core is inserted in a coil, what happens to the inductance?
   A. It increases.
   B. It decreases.
   C. It stays the same.
   D. It becomes voltage-dependent.
ANSWER A: Never adjust the iron core in a coil unless you know exactly what you are doing. They are usually factory pre-set, and wax-sealed so they seldom vibrate loose.

3AF-2-3.4 As a brass core is inserted in a coil, what happens to the inductance?
   A. It increases.
   B. It decreases.
   C. It stays the same.
   D. It becomes voltage-dependent.
ANSWER B: You can spot a coil with a brass core because it is usually larger and a small jewelers screwdriver fits the slot perfectly. Brass has the opposite effect of iron—the more brass you add in the core, the less the inductance.

3AF-2-4.1 What is the symbol used to represent an adjustable inductor on schematic diagrams?

A.   B.   C.   D.

ANSWER A: The arrow indicates the inductor is adjustable.

3AF-2-4.2 What is the symbol used to represent an iron-core inductor on schematic diagrams?

A.   B.   C.   D.

ANSWER B: The two lines above the coil indicate an iron core. Don't get the coil symbol mixed up with the resistor symbol.

3AF-2-4.3 What is the symbol used to represent an inductor wound over a toroidal core on schematic diagrams?

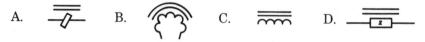

A.   B.   C.   D.

ANSWER C: There is no difference in the way a toroidal core is shown on a schematic diagram.

3AF-3-1.1 What is a capacitor dielectric?
A. The insulating material used for the plates
B. The conducting material used between the plates
C. The ferrite material that the plates are mounted on
D. The insulating material between the plates

ANSWER D: In paper capacitors, the dielectric is actually paper treated to make it a good insulator.

3AF-3-1.2 What are the component parts of a capacitor?
A. Two or more conductive plates with an insulating material between them
B. The wire used in the winding and the core material
C. Two or more layers of silicon material
D. Two insulating plates with a conductive material between them

ANSWER A: The plates never touch.

3AF-3-1.3 What is an electrolytic capacitor?
A. A capacitor whose plates are formed on a thin ceramic layer
B. A capacitor whose plates are separated by a thin strip of mica insulation
C. A capacitor whose dielectric is formed on one set of plates through electrochemical action
D. A capacitor whose value varies with applied voltage

ANSWER C: This type of capacitor offers large amounts of capacity in a small package. The plates are separated by a paste called the dielectric. Some electrolytic capacitors are polarized so that the end marked with a + must be connected to the positive side of a circuit.

3AF-3-1.4 What is a paper capacitor?
A. A capacitor whose plates are formed on a thin ceramic layer
B. A capacitor whose plates are separated by a thin strip of mica insulation
C. A capacitor whose plates are separated by a layer of paper
D. A capacitor whose dielectric is formed on one set of plates through electrochemical action

ANSWER C: The paper capacitor does not have a positive or negative end. Very thin paper separates the plates.

3AF-3-2.1 What is a capacitor?
A. An electronic component that stores energy in a magnetic field
B. An electronic component that stores energy in an electric field
C. An electronic component that converts a high voltage to a lower voltage
D. An electronic component that converts power into heat

ANSWER B: Capacitors store their energy in an electric field, not a magnetic field, as in a coil.

3AF-3-2.2 What are the electrical properties of a capacitor?
   A. A capacitor stores a charge electrochemically and opposes a change in current.
   B. A capacitor stores a charge electromagnetically and opposes a change in current.
   C. A capacitor stores a charge electromechanically and opposes a change in voltage.
   D. A capacitor stores a charge electrostatically and opposes a change in voltage.

ANSWER D: The key term to look for is "statically."

3AF-3-2.3 What factors must be considered when selecting a capacitor for a circuit?
   A. Type of capacitor, capacitance and voltage rating
   B. Type of capacitor, capacitance and the kilowatt-hour rating
   C. The amount of capacitance, the temperature coefficient and the KVA rating
   D. The type of capacitor, the microscopy coefficient and the temperature coefficient

ANSWER A: Be careful when experimenting with capacitors—if you exceed their voltage rating, they can explode and cause physical damage.

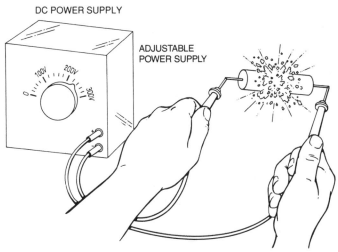

Overvoltage Causes Capacitor to Explode

3AF-3-2.4 How are the characteristics of a capacitor usually specified?
   A. In volts and amperes
   B. In microfarads and volts
   C. In ohms and watts
   D. In millihenrys and amperes

ANSWER B: Sometimes it's hard to identify small capacitors because they don't have much room for printing their values. Always keep your capacitors in well-marked bins, separated by microfarads and volts.

3AF-3-3.1 What factors determine the amount of capacitance in a capacitor?

A. The dielectric constant of the material between the plates, the area of one side of one plate, the separation between the plates and the number of plates

B. The dielectric constant of the material between the plates, the number of plates and the diameter of the leads connected to the plates

C. The number of plates, the spacing between the plates and whether the dielectric material is N type or P type

D. The dielectric constant of the material between the plates, the surface area of one side of one plate, the number of plates and the type of material used for the protective coating

ANSWER A: Memorize this answer for the examination, and read over the incorrect answers and see why they are wrong. Larger area, larger number of plates, closer spacing, and higher dielectric constant all increase capacitance.

3AF-3-3.2 As the plate area of a capacitor is increased, what happens to its capacitance?

A. Decreases
B. Increases
C. Stays the same
D. Becomes voltage dependent

ANSWER B: This question asks about "plate area," and the bigger the plates, the greater the capacitance.

3AF-3-3.3 As the plate spacing of a capacitor is increased, what happens to its capacitance?

A. Increases
B. Stays the same
C. Becomes voltage dependent
D. Decreases

ANSWER D: This question is dramatically different than the one above. Here they are asking about plate "spacing." The further apart the plates, the less capacitance.

3AF-3-4.1 What is the symbol used to represent an electrolytic capacitor on schematic diagrams?

A.   B.   C.   D.

ANSWER D: Notice that one of the vertical lines in a capacitor symbol is a convex line, not a straight line as in the battery symbol shown in "C."

3AF-3-4.2 What is the symbol used to represent a variable capacitor on schematic diagrams?

A.   B.   C.   D.

ANSWER A: A variable capacitor symbol has an arrow through it.

Subelement 3AG – Practical Circuits (1 examination question from 17 questions in 3AG)

3AG-1-1.1 Which frequencies are attenuated by a low-pass filter?
A. Those above its cut-off frequency
B. Those within its cut-off frequency
C. Those within 50 kHz on either side of its cut-off frequency
D. Those below its cut-off frequency

ANSWER A: The word "attenuated" means to reduce. Since a low-pass filter passes low frequencies, it attenuates all those frequencies above its cut-off frequency. See figure at question 3AG-1-1.2

3AG-1-1.2 What circuit passes electrical energy below a certain frequency and blocks electrical energy above that frequency?
A. A band-pass filter
B. A high-pass filter
C. An input filter
D. A low-pass filter

ANSWER D: Read this question carefully, too. Since it passes energy below a certain frequency, it must be a low-pass filter.

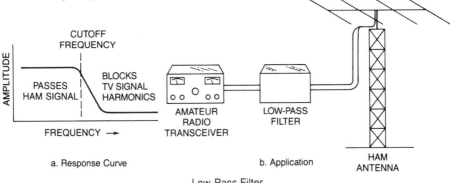

a. Response Curve   b. Application   HAM ANTENNA

Low-Pass Filter

3AG-1-2.1 Why does virtually every modern transmitter have a built-in low-pass filter connected to its output?
A. To attenuate frequencies below its cutoff point
B. To attenuate low frequency interference to other amateurs
C. To attenuate excess harmonic radiation
D. To attenuate excess fundamental radiation

ANSWER C: Your worldwide set already has a built-in low pass filter connected to its output. This usually is fine for reducing harmonics.

3AG-1-2.2 You believe that excess harmonic radiation from your transmitter is causing interference to your television receiver. What is one possible solution for this problem?
A. Install a low-pass filter on the television receiver
B. Install a low-pass filter at the transmitter output.
C. Install a high-pass filter on the transmitter output.
D. Install a band-pass filter on the television receiver.

ANSWER B: Install a low pass filter only on your worldwide radios. Don't install one on a VHF or UHF transceiver.

3AG-2-1.1 What circuit passes electrical energy above a certain frequency and attenuates electrical energy below that frequency?
A. A band-pass filter
B. A high-pass filter
C. An input filter
D. A low-pass filter

ANSWER B: Read this question carefully! Since it passes energy above a certain frequency, it must be a high-pass filter.

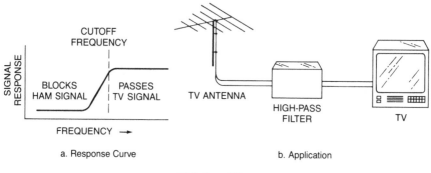

a. Response Curve

b. Application

High-Pass Filter

3AG-2-2.1 Where is the proper place to install a high-pass filter?
A. At the antenna terminals of a television receiver
B. Between a transmitter and a Transmatch
C. Between a Transmatch and the transmission line
D. On a transmitting antenna

ANSWER A: The best place to install the high-pass filter is right at the television's antenna jack. This helps filter out Amateur Radio high frequency signals that might travel down the twin lead or coax braid. A high-pass filter is not used on TVs that are connected to a subscription cable system—cable TV installations usually screen out all types of ham interference.

3AG-2-2.2 Your Amateur Radio transmissions cause interference to your television receiver even though you have installed a low-pass filter at the transmitter output. What is one possible solution for this problem?
A. Install a high-pass filter at the transmitter terminals.
B. Install a high-pass filter at the television antenna terminals.
C. Install a low-pass filter at the television antenna terminals also.
D. Install a band-pass filter at the television antenna terminals.

ANSWER B: If the television receiver is being operated off of a regular TV antenna, a high pass filter at the television antenna terminals may help. If it's a cable installation, use no high pass filter. Contact the cable company.

3AG-3-1.1 What circuit attenuates electrical energy above a certain frequency and below a lower frequency?
A. A band-pass filter
B. A high-pass filter
C. An input filter
D. A low-pass filter

ANSWER A: Read this question carefully, because it states that this particular filter blocks energy above, as well as below, a certain frequency. Therefore, this must be a band-pass filter.

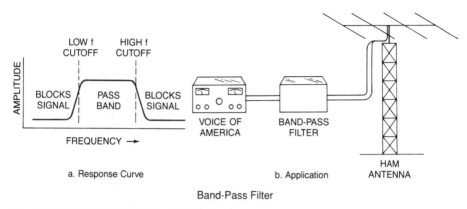

a. Response Curve   b. Application   HAM ANTENNA

Band-Pass Filter

3AG-3-1.2 What general range of RF energy does a band-pass filter reject?
   A. All frequencies above a specified frequency
   B. All frequencies below a specified frequency
   C. All frequencies above the upper limit of the band in question
   D. All frequencies above a specified frequency and below a lower specified frequency

ANSWER D: Most shortwave sets and scanner receivers employ automatic band-pass filters that are switched on in proper sequence when you change to a new megahertz frequency.

3AG-3-2.1 The IF stage of a communications receiver uses a filter with a peak response at the intermediate frequency. What term describes this filter response?
   A. A band-pass filter
   B. A high-pass filter
   C. An input filter
   D. A low-pass filter

ANSWER A: Band-pass filters are used in many types of radio receivers.

3AG-4-1.1 What circuit is likely to be found in all types of receivers?
   A. An audio filter
   B. A beat frequency oscillator
   C. A detector
   D. An RF amplifier

ANSWER C: All radio receivers need a circuit to demodulate or detect the impressed information on the rf carrier. A common detector detects the audio signal from the rf carrier.

3AG-4-1.2 What type of transmitter does this block diagram represent?
A. A simple packet-radio transmitter
B. A simple crystal-controlled transmitter
C. A single-sideband transmitter
D. A VFO-controlled transmitter

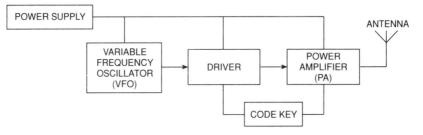

ANSWER D: Block diagrams are easy to identify—look at your possible answers, and then study the diagram for the key block elements. Since we see a complete transmitter setup along with the variable frequency oscillator, answer B is incorrect. Answer D is the correct answer. Always completely read all possible answer choices because one may be more correct than another one—as in this case.

3AG-4-1.3 What type of transmitter does this block diagram represent?
A. A simple packet-radio transmitter
B. A simple crystal-controlled transmitter
C. A single-sideband transmitter
D. A VFO-controlled transmitter

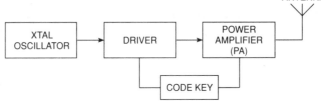

ANSWER B: On your Technician examination, they may show you a block diagram and want you to identify it. That's simple. If you spot a telegraph key, it must be a telegraphy transmitter. In this case, the crystal (XTAL) oscillator identifies the type of transmitter.

3AG-4-1.4 What is the unlabeled block (?) in this diagram?
A. An AGC circuit   C. A power supply
B. A detector   D. A VFO circuit

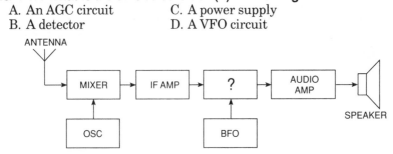

ANSWER B: The detector is needed to demodulate the audio from the carrier of mixed IF amplifier and beat frequency oscillator (BFO) output.

3AG-4-1.5 What type of device does this block diagram represent?
A. A double-conversion receiver
B. A variable-frequency oscillator
C. A simple superheterodyne receiver
D. A simple CW transmitter

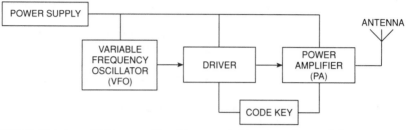

ANSWER D: This is a CW transmitter. You can tell by the telegraph key.

3AG-4-2.1 What type of device does this block diagram represent?
A. A double-conversion receiver
B. A variable-frequency oscillator
C. A simple superheterodyne receiver
D. A simple FM receiver

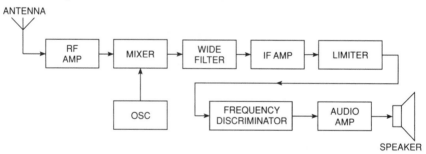

ANSWER D: The frequency discriminator gives us a clue that this is a simple FM receiver.

3AG-4-2.2 What is the unlabeled block (?) in this diagram?
A. A band-pass filter   C. A reactance modulator
B. A crystal oscillator   D. A rectifier modulator

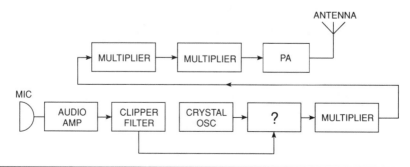

ANSWER C: All FM transmitters, as illustrated in this diagram, feature a reactance modulator.

Subelement 3AH – Signals and Emissions (2 examination questions from 28 questions in 3AH)

3AH-1.1 What is the meaning of the term *modulation?*

    A. The process of varying some characteristic of a carrier wave for the purpose of conveying information
    B. The process of recovering audio information from a received signal
    C. The process of increasing the average power of a single-sideband transmission
    D. The process of suppressing the carrier in a single-sideband transmitter

ANSWER A: When we modulate a transmitter, we usually impress our voice or audio tones on the carrier wave.

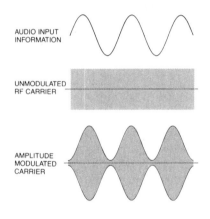

AUDIO INPUT INFORMATION

UNMODULATED RF CARRIER

AMPLITUDE MODULATED CARRIER

Amplitude Modulation

3AH-2-1.1 If the modulator circuit of your FM transmitter fails, what emission type would likely result?

    A. An unmodulated carrier wave
    B. A phase-modulated carrier wave
    C. An amplitude-modulated carrier wave
    D. A frequency-modulated carrier wave

ANSWER A: If there is no modulation circuit in an FM transmitter operating properly, there would be an unmodulated carrier wave.

3AH-2-1.2 What emission does not have sidebands resulting from modulation?

    A. AM phone
    B. Test
    C. FM Phone
    D. RTTY

ANSWER B: In the "test" mode, an unmodulated carrier wave would have no sidebands.

3AH-2-2.1 What is the FCC emission designator for a Morse code telegraphy signal produced by switching the transmitter output on and off?
 A. Test
 B. AM Phone
 C. CW
 D. RTTY
ANSWER C: Morse code is a form of continuous wave, or CW.

3AH-2-2.2 What is CW?
 A. Morse code telegraphy using amplitude modulation
 B. Morse code telegraphy using frequency modulation
 C. Morse code telegraphy using phase modulation
 D. Morse code telegraphy using pulse modulation
ANSWER A: Simple CW.

3AH-2-3.1 What is RTTY?
 A. Amplitude-keyed telegraphy
 B. Frequency-shift-keyed telegraphy
 C. Frequency-modulated telephony
 D. Phase-modulated telephony
ANSWER B: FSK is the same as RTTY on worldwide frequencies.

3AH-2-3.2 What is the emission designation for telegraphy by frequency shift keying without the use of a modulating tone?
 A. RTTY
 B. MCW
 C. CW
 D. Sideband phone
ANSWER A: On the worldwide bands, we shift the frequency of a pure carrier wave back and forth to create a modulating tone in the receiver. This is called RTTY, or radioteleprinter.

3AH-2-4.1 What emission type results when an on/off keyed audio tone is applied to the microphone input of an FM transmitter?
 A. RTTY
 B. MCW
 C. CW
 D. Sideband phone
ANSWER B: There is not a lot of CW around using an FM transmitter and a tone on FM frequencies; however, if you do tune in an FM code practice, it is called "modulated CW," or MCW.

3AH-2-4.2 What is tone-modulated international Morse code telegraphy?
 A. Telephony produced by audio fed into an FM transmitter
 B. Telegraphy produced by an on/off keyed audio tone fed into a CW transmitter
 C. Telegraphy produced by on/off keying of the carrier amplitude
 D. Telegraphy produced by an on/off keyed audio tone fed into an FM transmitter

ANSWER D: It would be impractical to send telegraphy using the touch-tone pad of your handie-talkie. It can be done, but don't do it—it gets your transmitter hot and consumes a lot of battery power.

3AH-2-5.1 What is the emission designated as "MCW"?
A. Frequency-modulated telegraphy using audio tones
B. Frequency-modulated telephony
C. Frequency-modulated facsimile using audio tones
D. Phase-modulated television

ANSWER A: MCW stands for modulated carrier wave, and this is what we use for code practice on FM when using audio tones.

3AH-2-5.2 In an ITU emission designator like A1A, what does the first symbol describe?
A. The nature of the signal modulating the main carrier
B. The type of the information to be transmitted
C. The speed of a radiotelegraph transmission
D. The type of modulation of the main carrier

ANSWER D: The first letter indicates the type of modulation to be transmitted.

3AH-2-5.3 What emission type results when an on-off keyed audio oscillator is connected to the microphone jack of an FM phone transmitter?
A. SS
B. RTTY
C. MCW
D. Image

ANSWER C: This is Morse code practice using an FM transmitter

3AH-2-6.1 In an ITU emission designator like F3B, what does the second symbol describe?
A. The nature of the signal modulating the main carrier
B. Tye type of modulation of the main carrier
C. The type of information to be transmitted
D. The frequency modulation index of a carrier

ANSWER A: The second symbol describes the nature of the signal modulating the main carrier that is being sent. See Table 2A-17.2.

3AH-2-6.2 How would you transmit packet using an FM 2-meter transceiver?
A. Use your telegraph key to interrupt the carrier wave
B. Modulate your FM transmitter with audio tones from a terminal node controller
C. Use your mike for telephony
D. Use your touch-tone (DTMF) key pad to signal in Morse code

ANSWER B: We modulate audio tones from a "terminal node controller" to create packet communications on FM.

3AH-2-7.1 What type of emission results when speaking into the micro-phone of a 2-meter FM handheld transceiver?
- A. Amplitude-modulated phone
- B. Code telegraphy
- C. An unmodulated carrier wave
- D. Frequency-modulated phone

ANSWER D: Remember the word "phone" when talking over a microphone. And since we are using FM, this is frequency-modulated phone.

3AH-2-7.2 What emission type do most 2-meter FM transmitters transmit?
- A. Interrupted pure carrier waves
- B. Frequency-modulated phone
- C. Single-sideband voice emissions
- D. Amplitude-modulated carrier waves

ANSWER B: FM is frequency modulation, widely used for telephony or "phone."

3AH-2-8.1 What is the most important consideration when installing a 10-meter dipole inside an attic?
- A. It will exhibit a low angle of radiation
- B. The dipole must always be run horizontally polarized
- C. It will be covered by an insulation to prevent fire and high enough to prevent touched during transmission
- D. Dipoles usually don't work in attics

ANSWER C: Make sure that any transmitting antenna in an attic is covered by an insulation to prevent fire or someone getting burned touching it. There are some excellent 10-meter vertical antennas with a high voltage insulation to specifically keep the antenna safe from fire, touching, or a brush with electric wires.

3AH-2-8.2 Which type of transmitter will produce a frequency modulated carrier wave?
- A. A CW transmitter
- B. An amplitude-modulation transmitter
- C. A single-sideband transmitter
- D. A phase-modulated transmitter

ANSWER D: A phase-modulated transmitter will also create frequency modulation. Just say "phase" "frequency," and they begin with the same sounding letter.

3AH-3.1 What is the term used to describe a constant-amplitude radio-frequency signal?
- A. An RF carrier
- B. An AF carrier
- C. A sideband carrier
- D. A subcarrier

ANSWER A: The carrier is a pure signal, without tone.

3AH-3.2 What is another name for an unmodulated radio-frequency signal?
- A. An AF carrier
- B. An RF carrier
- C. A sideband carrier
- D. A subcarrier

ANSWER B: Use an RF carrier to tune a worldwide transceiver.

3AH-4.1 What characteristic makes FM telephony especially well-suited for local VHF/UHF radio communications?
  A. Good audio fidelity and intelligibility under weak-signal conditions
  B. Better rejection of multipath distortion than the AM modes
  C. Good audio fidelity and high signal-to-noise ratio above a certain signal amplitude threshold
  D. Better carrier frequency stability than the AM modes
ANSWER C: FM is only good when there is a high signal-to-noise ratio. Since most FM work is line-of-sight, FM is a delightful mode to use on frequencies above 50 MHz.

3AH-5.1 What emission is produced by a transmitter using a reactance modulator?
  A. CW
  B. Unmodulated carrier
  C. Single-sideband, suppressed-carrier phone
  D. Phase-modulated phone
ANSWER D: Some VHF transceivers use phase modulation, as opposed to frequency modulation. They sound a little bit different on the air, and their characteristic is also different—G3E.

3AH-5.2 What other emission does phase modulation most resemble?
  A. Amplitude modulation
  B. Pulse modulation
  C. Frequency modulation
  D. Single-sideband modulation
ANSWER C: Frequency modulation and phase modulation sound almost the same unless you really listen carefully.

3AH-6.1 Many communications receivers have several IF filters that can be selected by the operator. Why do these filters have different bandwidths?
  A. Because some ham bands are wider than others
  B. Because different bandwidths help increase the receiver sensitivity
  C. Because different bandwidths improve S-meter readings
  D. Because some emission types occupy a wider frequency range than others
ANSWER D: A better worldwide transceiver has several different IF filters that may be selected. Look for this feature when selecting your worldwide base station.

3AH-6-1.2 List the following signals in order of increasing bandwidth (narrowest signal first): CW, FM voice, RTTY, SSB voice.
  A. RTTY, CW, SSB voice, FM voice
  B. CW, FM voice, RTTY, SSB voice
  C. CW, RTTY, SSB voice, FM voice
  D. CW, SSB voice, RTTY, FM voice
ANSWER C: This gives you a clue why FM voice frequencies are always found in the VHF and UHF spectrum—and the only place FM is found on worldwide frequencies. Because of its wide bandwidth, (between 29.5 and 29.7 MHz), using FM below these frequencies exceeds bandwidth limitations.

3AI–ANTENNAS AND FEED LINES 3

3AH-7-1.1 To what is the deviation of an FM transmission proportional?
A. Only the frequency of the audio modulating signal
B. The frequency and the amplitude of the audio modulating signal
C. The duty cycle of the audio modulating signal
D. Only the amplitude of the audio modulating signal

ANSWER D: The "deviation" is the swing of a frequency modulated signal. The swing is proportional to the amplitude of the audio signal. Less amplitude, less swing.

3AH-7-2.1 What is the result of overdeviation in an FM transmitter?
A. Increased transmitter power consumption
B. Out-of-channel emissions (splatter)
C. Increased transmitter range
D. Inadequate carrier suppression

ANSWER B: If you overdrive the audio stage of an FM transceiver, you will create over-deviation that creates interference to adjacent channel users.

3AH-7-2.2 What is splatter?
A. Interference to adjacent signals caused by excessive transmitter keying speeds
B. Interference to adjacent signals caused by improper transmitter neutralization
C. Interference to adjacent signals caused by overmodulation of a transmitter
D. Interference to adjacent signals caused by parasitic oscillations at the antenna

ANSWER C: Back off the microphone if someone indicates your signal is "splattering." On some FM sets, you can turn down the modulation gain in the mike circuit. On other sets, you simply turn down the amount of deviation on the deviation pot, which is inside the transmitter.

Subelement 3AI – Antennas and Feed Lines (3 examination questions from 42 questions in 3AI)

3AI-1-1.1 What antenna type best strengthens signals from a particular direction while attenuating those from other directions?
A. A beam antenna
B. An isotropic antenna
C. A monopole antenna
D. A vertical antenna

ANSWER A: A beam antenna is just like your television antenna. It picks up signals better in one direction than in any other. Yagi antennas are beam antennas. See figure in 3AI-1-1.2.

3AI-1-1.2 What is a directional antenna?
A. An antenna whose parasitic elements are all constructed to be directors
B. An antenna that radiates in direct line-of-sight propagation, but not skywave or skip propagation

C. An antenna permanently mounted so as to radiate in only one direction

D. An antenna that radiates more strongly in some directions than others

ANSWER D: Good examples of directional antennas are quads, delta loops, and the Yagi.

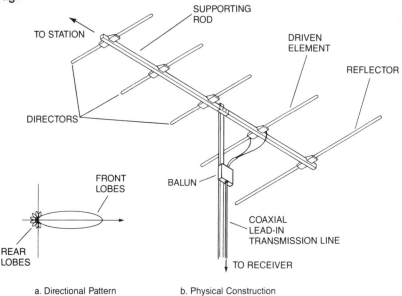

a. Directional Pattern   b. Physical Construction

A Beam Antenna — The Yagi Antenna
Source: *Antennas — Selection and Installation,* © 1986 Master Publishing, Inc., Richardson, Texas

3AI-1-1.3 What is a Yagi antenna?

A. Half-wavelength elements stacked vertically and excited in phase

B. Quarter-wavelength elements arranged horizontally and excited out of phase

C. Half-wavelength linear driven element(s) with parasitically excited parallel linear elements

D. Quarter-wavelength, triangular loop elements

ANSWER C: See figure in 3AI-1-1.2. The Yagi antenna is really a series of dipoles, all lined up to increase the signal gain in one direction. A dipole is usually a half-wavelength long. However, on the Yagi beam, some are longer and some are shorter than the driven dipole. The ones called reflectors are slightly longer than the driven element, and the ones called directors are slightly shorter. The combination makes the Yagi antenna very directional. Since the reflectors and directors are not electrically connected to the driven element, they are said to be "parasitically excited." Don't let that big long answer throw you!

3AI-1-1.4 What is the general configuration of the radiating elements of a horizontally-polarized Yagi?

A. Two or more straight, parallel elements arranged in the same horizontal plane

B. Vertically stacked square or circular loops arranged in parallel horizontal planes

    C. Two or more wire loops arranged in parallel vertical planes

    D. A vertical radiator arranged in the center of an effective RF ground plane

ANSWER A: The more directing elements a Yagi has, the greater its directivity. Most Yagis use only one reflector, but some may have two or three. See figure in 3AI-1-1.2.

3AI-1-1.5 What type of parasitic beam antenna uses two or more straight metal-tubing elements arranged physically parallel to each other?

    A. A delta loop antenna

    B. A quad antenna

    C. A Yagi antenna

    D. A Zepp antenna

ANSWER C: A Yagi antenna is a beam antenna. See figure in 3AI-1-1.2.

3AI-1-1.6 How many directly-driven elements does a Yagi antenna have?

    A. None; they are all parasitic

    B. One

    C. Two

    D. All elements are directly driven

ANSWER B: For the examination, consider only one element as a driven element. However, in the real world of Amateur Radio, there are several beam manufacturers that drive more than one element. See figure in 3AI-1-1.2.

3AI-1-1.7 What is a parasitic beam antenna?

    A. An antenna where the director and reflector elements receive their RF excitation by induction or radiation from the driven element

    B. An antenna where wave traps are used to assure magnetic coupling among the elements

    C. An antenna where all elements are driven by direct connection to the feed line

    D. An antenna where the driven element receives its RF excitation by induction or radiation from the directors

ANSWER A: A beam antenna is an inexpensive investment for excellent power output and extraordinarily good receiving capabilities. It's much better to buy a beam antenna than a power amplifier.

3AI-1-2.1 What is a cubical quad antenna?

    A. Four parallel metal tubes, each approximately 1/2 electrical wavelength long

    B. Two or more parallel four-sided wire loops, each approximately one electrical wavelength long

    C. A vertical conductor 1/4 electrical wavelength high, fed at the bottom

    D. A center-fed wire 1/2 electrical wavelength long

ANSWER B: You can add reflectors and directors to your quad antenna system to give it some real punch in just one direction. Just like the Yagi, the reflector element is slightly larger, and the director loop is slightly smaller. See 3AI-1-2.2.

3AI-1-2.2 What kind of antenna array is composed of a square full-wave closed loop driven element with parallel parasitic element(s)?
A. Delta loop
B. Cubical quad
C. Dual rhombic
D. Stacked Yagi

ANSWER B: If you have enough room, and understanding neighbors, a cubical quad antenna is a good monster antenna to put up. It forms a cube that is one full wavelength long when erected.

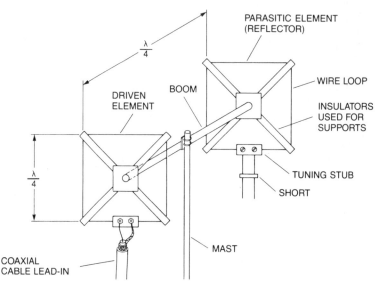

A Two-Element Quad Antenna — Horizontally Polarized
Source: *Antennas — Selection and Installation,* © 1986 Master Publishing, Inc., Richardson, Texas

3AI-1-2.3 Approximately how long is one side of the driven element of a cubical quad antenna?
A. 2 electrical wavelengths
B. 1 electrical wavelength
C. 1/2 electrical wavelength
D. 1/4 electrical wavelength

ANSWER D: Since the entire quad is one wavelength, each side must be one-quarter wavelength.

3AI-1-2.4 Approximately how long is the wire in the driven element of a cubical quad antenna?
A. 1/4 electrical wavelength
B. 1/2 electrical wavelength
C. 1 electrical wavelength
D. 2 electrical wavelengths

ANSWER C: Read this question carefully—it doesn't ask about one side, but rather how long is the entire wire in the driven element of a quad. One electrical wavelength. See figure in 3AI-1-2.2.

3AI-1-3.1 What is a delta loop antenna?
A. A variation of the cubical quad antenna, with triangular elements
B. A large copper ring, used in direction finding
C. An antenna system composed of three vertical antennas, arranged in a triangular shape
D. An antenna made from several coils of wire on an insulating form

ANSWER A: The delta loop is a great one to string in trees. It's one complete wavelength, with each side one-third wavelength long.

3AI-2-1.1 To what does the term horizontal as applied to wave polarization refer?
A. The magnetic lines of force in the radio wave are parallel to the earth's surface.
B. The electric lines of force in the radio wave are parallel to the earth's surface.
C. The electric lines of force in the radio wave are perpendicular to the earth's surface.
D. The radio wave will leave the antenna and radiate horizontally to the destination.

ANSWER B: The key here is "electric lines." Make sure that your antenna, for local VHF work, is polarized in the same plane as your friend's. Almost all repeaters use vertical polarization.

3AI-2-1.2 What electromagnetic wave polarization does a cubical quad antenna have when the feed point is in the center of a horizontal side?
A. Circular
B. Helical
C. Horizontal
D. Vertical

ANSWER C: Feed any quad antenna at the center of a horizontal side, and its polarization will be horizontal.

3AI-2-1.3 What electromagnetic wave polarization does a cubical quad antenna have when all sides are at 45 degrees to the earth's surface and the feed point is at the bottom corner?
A. Circular
B. Helical
C. Horizontal
D. Vertical

ANSWER C: Since the quad is being fed at the bottom, even though it is a diamond-shaped quad, any bottom feed is horizontal.

3AI-2-2.1 What is the polarization of electromagnetic waves radiated from a half-wavelength antenna perpendicular to the earth's surface?
A. Circularly polarized waves
B. Horizontally polarized waves
C. Parabolically polarized waves
D. Vertically polarized waves

ANSWER D: Perpendicular to the earth's surface means vertical and a vertical antenna radiates vertically polarized electromagnetic waves.

3AI-2-2.2 What is the electromagnetic wave polarization of most man-made electrical noise radiation in the HF-VHF spectrum?
A. Horizontal
B. Left-hand circular
C. Right-hand circular
D. Vertical

ANSWER D: Think of this answer as just backwards to what you might think when looking at your local horizontal power lines. Those horizontal power lines emit vertical electromagnetic waves. Most electrical interference is loudest when it is vertically polarized.

3AI-2-2.3 To what does the term vertical as applied to wave polarization refer?
A. The electric lines of force in the radio wave are parallel to the earth's surface.
B. The magnetic lines of force in the radio wave are perpendicular to the earth's surface.
C. The electric lines of force in the radio wave are perpendicular to the earth's surface.
D. The radio wave will leave the antenna and radiate vertically into the ionosphere.

ANSWER C: For worldwide communications, polarization is not critical because skywave refraction will turn them every which way. However, for VHF propagation, almost everyone uses vertical polarization for best base to mobile range.

3AI-2-2.4 What electromagnetic wave polarization does a cubical quad antenna have when the feed point is in the center of a vertical side?
A. Circular
B. Helical
C. Horizontal
D. Vertical

ANSWER D: Feed any quad on a vertical side, and its polarization will be vertical.

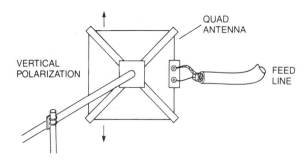

Vertical Polarization of Quad Antenna Fed on Vertical Side

3AI-2-2.5 What electromagnetic wave polarization does a cubical quad antenna have when all sides are at 45 degrees to the earth's surface and the feed point is at a side corner?
A. Circular
B. Helical
C. Horizontal
D. Vertical
ANSWER D: Since the quad is being fed on the side, any side-feed is vertical.

3AI-3-1.1 What is meant by the term *standing wave ratio*?
A. The ratio of maximum to minimum inductances on a feed line
B. The ratio of maximum to minimum resistances on a feed line
C. The ratio of maximum to minimum impedances on a feed line
D. The ratio of maximum to minimum voltages on a feed line
ANSWER D: Standing wave ratio (SWR) is the ratio of maximum voltage to minimum voltage along a transmission line. It is also a ratio of the maximum current to minimum current along a transmission line. It also is the ratio of the power fed forward along a transmission line to the power reflected back along a transmission line.

3AI-3-1.2 What is standing wave ratio a measure of?
A. The ratio of maximum to minimum voltage on a feed line
B. The ratio of maximum to minimum reactance on a feed line
C. The ratio of maximum to minimum resistance on a feed line
D. The ratio of maximum to minimum sidebands on a feed line
ANSWER A: SWR is always a ratio, either of maximum to minimum voltages, or maximum to minimum currents, or forward to reflected power, on a transmission line.

3AI-3-2.1 What is meant by the term *forward power*?
A. The power traveling from the transmitter to the antenna
B. The power radiated from the front of a directional antenna
C. The power produced during the positive half of the RF cycle
D. The power used to drive a linear amplifier
ANSWER A: You should strive for maximum forward power and minimum reflected power. This will give you a perfect match.

3AI-3-2.2 What is meant by the term *reflected power*?
A. The power radiated from the back of a directional antenna
B. The power returned to the transmitter from the antenna
C. The power produced during the negative half of the RF cycle
D. Power reflected to the transmitter site by buildings and trees
ANSWER B: If you have high amounts of reflected power, better check out what's happening at the antenna. Your antenna is either cut to the wrong frequency or has an impedance mismatch. Hopefully, your reflected power will always be less than a watt.

3AI-3-3.1 What happens to the power loss in an unbalanced feed line as the standing wave ratio increases?
A. It is unpredictable.
B. It becomes nonexistent.
C. It decreases.
D. It increases.
ANSWER D: As the SWR increases, so does power loss in coax cable.

3AI-3-3.2 What type of feed line is best suited to operating at a high standing wave ratio?
A. Coaxial cable
B. Flat ribbon "twin lead"
C. Parallel open-wire line
D. Twisted pair
ANSWER C: Hams never like to operate their equipment with a high SWR, but if they must, parallel feed lines are less apt to arc over.

3AI-3-3.3 What happens to RF energy not delivered to the antenna by a lossy coaxial cable?
A. It is radiated by the feed line.
B. It is returned to the transmitter's chassis ground.
C. Some of it is dissipated as heat in the conductors and dielectric.
D. It is canceled because of the voltage ratio of forward power to reflected power in the feed line.
ANSWER C: Experienced hams can tell how good an antenna match is by feeling the coaxial cable feed line after a few minutes of transmitting. A warm feed line indicates high SWR. If your transmitter is getting red hot, that's also an indication of high SWR. This is a condition that needs to be corrected before melt-down.

3AI-4-1.1 What is a balanced line?
A. Feed line with one conductor connected to ground
B. Feed line with both conductors connected to ground to balance out harmonics
C. Feed line with the outer conductor connected to ground at even intervals
D. Feed line with neither conductor connected to ground
ANSWER D: Balanced line is similar to television twin lead. Two parallel conductors are separated by a dielectric. For Amateur Radio use, hams employ plastic spacers that keep balanced line wires several inches apart. This gives a usual impedance of about 600 ohms. Neither conductor is grounded—both are hot.

3AI-4-1.2 What is an unbalanced line?
A. Feed line with neither conductor connected to ground
B. Feed line with both conductors connected to ground to suppress harmonics
C. Feed line with one conductor connected to ground
D. Feed line with the outer conductor connected to ground at uneven intervals
ANSWER C: Coaxial cable is an unbalanced feed line. Your radio waves travel along the center conductor. The grounded outside braid keeps your radio waves in, and interference out.

3AI-4-2.1 What is a balanced antenna?
- A. A symmetrical antenna with one side of the feed point connected to ground
- B. An antenna (or a driven element in an array) that is symmetrical about the feed point
- C. A symmetrical antenna with both sides of the feed point connected to ground, to balance out harmonics
- D. An antenna designed to be mounted in the center

ANSWER B: A dipole and a Yagi are good examples of a balanced antenna. Neither side is connected to ground.

3AI-4-2.2 What is an unbalanced antenna?
- A. An antenna (or a driven element in an array) that is not symmetrical about the feed point
- B. A symmetrical antenna, having neither half connected to ground
- C. An antenna (or a driven element in an array) that is symmetrical about the feed point
- D. A symmetrical antenna with both halves coupled to ground at uneven intervals

ANSWER A: Most vertical antennas are considered "unbalanced." Half of the system is grounded. The ground could be the automobile chassis, a sea water ground, or quarter-wavelength ground radials run on your rooftop. The better the ground, the better the signal.

3AI-4-3.1 What device can be installed on a balanced antenna so that it can be fed through a coaxial cable?
- A. A balun
- B. A loading coil
- C. A triaxial transformer
- D. A wavetrap

ANSWER A: Since coax is unbalanced, and we want to feed a balanced beam or dipole with coax cable, a transformer must be used to take the unbalanced signal and put it on both sides of the balanced antenna system. This device may be constructed, or purchased. It's simply called a balun—*balanced to unbalanced*—transformer.

3AI-4-3.2 What is a balun?
- A. A device that can be used to convert an antenna designed to be fed at the center so that it may be fed at one end
- B. A device that may be installed on a balanced antenna so that it may be fed with unbalanced feed line
- C. A device that can be installed on an antenna to produce horizontally polarized or vertically polarized waves
- D. A device used to allow an antenna to operate on more than one band

ANSWER B: Many directional antennas are shipped with a balun already installed at the feed point. This allows an immediate hook up of the antenna to coax cable.

3AI-5-1.1 List the following types of feed line in order of increasing attenuation per 100 feet of line (list the line with the lowest attenuation first): RG-8, RG-58, RG-174 and open-wire line.
  A. RG-174, RG-58, RG-8, open-wire line
  B. RG-8, open-wire line, RG-58, RG-174
  C. open-wire line, RG-8, RG-58, RG-174
  D. open-wire line, RG-174, RG-58, RG-8
ANSWER C: Open wire line has the least loss. However, it requires matching networks at the transceiver end of the circuit.

3AI-5-1.2 You have installed a tower 150 feet from your radio shack, and have a 6-meter Yagi antenna on top. Which of the following feed lines should you choose to feed this antenna: RG-8, RG-58, RG-59 or RG-174?
  A. RG-8
  B. RG-58
  C. RG-59
  D. RG-174
ANSWER A: RG-8 is the best choice out of the given answers. However, the author recommends RG-213 (a better grade of RG-8) or Belden #9913, which is the ultimate in low-loss, semi-flexible cable.

3AI-5-2.1 You have a 200-foot coil of RG-58 coaxial cable attached to your antenna, but the antenna is only 50 feet from your radio. To minimize feed-line loss, what should you do with the excess cable?
  A. Cut off the excess cable to an even number of wavelengths long.
  B. Cut off the excess cable to an odd number of wavelengths long.
  C. Cut off the excess cable.
  D. Roll the excess cable into a coil a tenth of a wavelength in diameter.
ANSWER C: Don't use any more cable than you must. Keep your cable lengths short.

3AI-5-2.2 How does feed-line length affect signal loss?
  A. The length has no effect on signal loss.
  B. As length increases, signal loss increases.
  C. As length decreases, signal loss increases.
  D. The length is inversly proportional to signal loss.
ANSWER B: The longer your cable runs, the greater the line loss.

3AI-5-3.1 What is the general relationship between frequencies passing through a feed line and the losses in the feed line?
  A. Loss is independent of frequency.
  B. Loss increases with increasing frequency.
  C. Loss decreases with increasing frequency.
  D. There is no predictable relationship.
ANSWER B: In all feed lines, the higher the frequency, the higher the losses for 100 feet. This is why it's important to use only low-loss coax at VHF and UHF frequencies.

3AI-5-3.2 As the operating frequency decreases, what happens to conductor losses in a feed line?
A. The losses decrease.
B. The losses increase.
C. The losses remain the same.
D. The losses become infinite.
ANSWER A: As frequency decreases, such as for worldwide use, coax cable losses decrease.

3AI-5-3.3 As the operating frequency increases, what happens to conductor losses in a feed line?
A. The losses decrease.
B. The losses increase.
C. The losses remain the same.
D. The losses decrease to zero.
ANSWER B: As frequency increases, coax cable feed-line losses increase.

3AI-6-1.1 You are using open-wire feed line in your amateur station. Why should you ensure that no one can come in contact with the feed line while you are transmitting?
A. Because contact with the feed line while tranmitting will cause a short circuit, probably damaging your transmitter
B. Because the wire is so small they may break it
C. Because contact with the feed line while transmitting will cause parasitic radiation
D. Because high RF voltages can be present on open-wire feed line
ANSWER D: Make sure no one can touch your open-wire feed lines. Both conductors have dangerous high voltages that will cause a nasty burn.

3AI-6-2.1 How can you minimize exposure to radio frequency energy from your transmitting antennas?
A. Use vertical polarization.
B. Use horizontal polarization.
C. Mount the antennas where no one can come near them.
D. Mount the antenna close to the ground.
ANSWER C: Since RF burns can be painful, make sure no one can accidentally touch your antenna.

At this point, go back and check your work. Put a check mark beside each question and answer you have memorized correctly. Continue to review the questions and answers with which you may be having a little problem. Let a friend read you the correct answer, and see if you can recite the question! Continue to review the questions and answers until you have mastered them so you don't miss any more than 1 in 10. Then get yourself scheduled and go down and take that Technician Class examination. The next chapter tells you how to get scheduled and what to expect when you get there.

Taking the Technician Class Examination

ABOUT THIS CHAPTER

Now that you have studied Element 2 and Element 3A question pools, you are ready for the big test for your Technician Class license. And if you already know the code, take the Element 1A code test and end up with a Technician Plus license.

This chapter tells you how the examination will be given, who is qualified to give it, and what happens after you complete the tests.

EXAMINATION ADMINISTRATION

For Technician Class No-Code License

Unlike the Novice Class that can be given anywhere by any two licensed General Class and higher hams at anytime, the Technician Class examination *must* be taken at a volunteer examination official test session.

The FCC no longer conducts Amateur Radio services examinations; all examinations are conducted by volunteer amateur operators. The test sessions are coordinated by national or regional volunteer-examiner coordinators (VECs) who accredit Extra Class and Advanced Class ham operators to serve as volunteer examiners (VEs).

Three officially certified VEs are required to administer a Technician Class examination. The VEs are not compensated for their time and skills, but they are permitted to charge you a test fee for certain reimbursable expenses incurred in preparing, processing, or administering the examination. The maximum fee is adjusted annually by the FCC. It is usually less than $5.25.

The three volunteer examiners form a team called a volunteer examination team (VET), and they offer local Technician Class examinations regularly at local sites to serve their community. The VETs generally coordinate closely and rotate examination sites so you should be able to find a test site near you. You can obtain information about VETs and examination sessions by checking with your local radio club, ham radio store, and local packet bulletin boards. Also, the local Radio Shack manager can often prove helpful here. If this is not convenient, write or call the VEC that serves your area. The VEC will be able to give you the name, address and telephone number of your local VET. A list of VECs that was current at the time of publication is given in the appendix on page 208.

Once you have found your examination location and set the possible date for the test session, contact the local VET examiners and pre-register for taking your examination. They will hold a seat for you at the next available session, and they will appreciate you contacting them ahead of time making a reservation. Don't be a no-show, and don't be a surprise-show. Call them ahead of time!

For Technician Plus Class License

With Technician License

Your examination will be administered just as for the Technician Class license except that you will only be signing up for an Element 1A 5-wpm Morse code test.

With Novice License

Your examination will be administered just as for the Technician Class license except that you will only be signing up for an Element 3A written examination. See Appendix for instructions for applicants who want to pursue Novice license before Technician Plus.

With a Knowledge of Code and No Operator License

Your examination will be administered just as for the Technician Class license except that you will be signing up for the Element 2 written examination, the Element 3A written examination, and the Element 1A 5-wpm code test.

Special Cases

Some applicants may have chosen to start up the Novice Class, have passed the Element 2 written examination, but have not passed the Element 1A 5-wpm code test. If it has not been a year since your Element 2 written examination was passed, you should sign up for a Element 3A written examination with the VET examiners. Your administration will be the same as for a Technician Class license. Bring your Form 610 to the test session to assure that you get credit for your passed Element 2 examination. You will be awarded a Technician Class operator's license when you pass.

If you are an applicant that has passed the Element 1A code test and a year has not passed, but have not passed the Element 2 or Element 3A written examinations, you should sign up for the two written examinations. Your examinations will be administered just as for the Technician Class, but you will be awarded a Technician Plus operator's license if you pass your examinations.

EXAM CONTENT

The FCC previously handled amateur services testing. They developed the questions, the multiple choices, and identified the correct answers. And it was all sort of secret! Neither the questions, nor the correct answers, were really widely known. That has changed. The FCC has been transferring testing responsibility—including development of examination questions—over to the amateur community in phases since 1982, when President Reagan signed legislation providing for volunteer amateur operator examinations above the Novice Class. VECs periodically revise the questions in the various amateur operator class question pools, and recommend multiple-choice answers to the VEs. The VEs are responsible for the answers, but most VEs accept the multiple-choice answers, both correct and incorrect, supplied by the VECs.

What most VECs and VEs have adopted is an amateur services examination system similar to that used by the FAA for testing pilots. Pilots know all the possible questions that might be on an examination. Amateur operators are given that same privilege with their questions when they take an examination. If it works for the FAA, it should work for the FCC!

COMPLETING THE FCC FORM 610

A FCC Form 610, *Application For Amateur Radio Station And / Or Operator License,* is included with this book. You should complete it as best you can before going to the examination site.

Complete the front side of the form, answering questions 2C, 5, 6, 7, 8, 9, 10, and signing on Line 13 and dating in Box 14. Don't make any marks at the top of the form, nor do anything on the back of the form—this is for your volunteer examiners.

Don't worry about the expiration date in the upper right-hand corner of your 610 form. This date pertains to the printing of the form, not necessarily its use. As long as you have the form from this book, it will be up to date.

You may obtain additional Form 610s from your area VEC, or directly from the FCC, Forms Distribution Center, 2803 52nd Avenue, Hyattsville, MD 20781, 202/632-3676. Tell them how many forms and where to send them. If you have any questions about how to fill out the form, you may contact the FCC at 717/337-1212.

Look over the sample 610 we have filled out in this book, and make sure you didn't leave out any information. We recommend you use your home address as your station location, even though you will be operating your equipment from portable locations, in a car, or out on a boat.

Be sure your signature on Line 13 matches your name on Line 5, and make sure they can read your printing!

Your Examiner's Portion

The VET (or your two ham buddies, if through Novice path) will complete the administering VEs' report on the front and back of the FCC Form 610, and send it in to the proper agency for processing. For new applications, it takes approximately 6 to 8 weeks to finally receive your official FCC call signs. Your call letters will be sent directly to the mailing address you put down on your Form 610.

SPECIAL PROVISIONS FOR THE HANDICAPPED

The amateur service welcomes handicapped operators. The FCC rules encourage examiners to take any necessary steps to assist the handicapped operator through the examination process. For the visually impaired, the examiners could read you the questions and ask for spoken answers. If you're not able to write clearly, the examiners will assist you in transcribing the answers onto the answer sheet. However, I would encourage you to document your physical handicap with correspondence from your physician so that examiners will know how they might assist you in getting through the testing process. This procedure protects the VEs from unwarranted criticism of their judgement to extend special privileges.

Be sure to tell the VET ahead of time that you may require a special examination for the handicapped—as long as they know that you are coming with special circumstances, they will work closely with you to help you pass the test. Handicapped operators are a valuable resource of quality trained ham radio operators. Handicapped code test exemption forms are available from your local VET for 13-wpm and 20-wpm code tests only.

TAKING THE EXAMINATION

Get a good night's sleep before exam day. Continue to study both theory elements up to the moment you go into the room. Gordon West Radio School offers cassette tapes that cover the theory needed for the Element 2 and Element 3A written examinations. You may want to purchase a set and listen to them while you are waiting to take the test. They also have upgrade theory and code cassettes for all classes of amateur operator licenses.

What to Bring to the Exam

Here's what you'll need for your Technician Class written examination (or optional code test, if you are being tested for Technician Plus):
1. Examination fee of approximately $5.50 in cash. (If you happen to be going for the Novice license, no fee is required.)
2. Personal identification with a photo.
3. Properly filled out FCC Form 610.

4. Any other certificate of completions or signed 610 forms from other examiners who have tested you within 365 days of this test session date.
5. Some sharp pencils and fine-tip pens. Bring a backup!
6. Calculators may be used, so bring your calculator.
7. A letter from your physician indicating you are handicapped if you are requesting a special handicapped exam. If you need any special equipment, such as a braille writer, you may supply it.
8. Any other item that the VET asks you to bring. I suggest donuts— they are always welcome. Remember, these volunteer examiners receive no pay for their work.

Check and Double-Check

Read over the examination questions carefully. Take your time in looking for the correct answer. Some answers start out looking correct, but end up wrong. Don't speed read the test.

When you are finished with an examination, go back over every question and double-check your answers. Try a game where you read what you have selected as the correct answer, and see if it agrees with the question.

When the examiners hand out the examination material, put your name, date, and test number on the answer sheet. *Make No Marks On the Multiple-Choice Question Sheet.* Only write on the answer sheets.

When you are finished with the examination, turn in all of your paperwork. Tell the examiners how much you appreciate their unselfish efforts to help promote ham radio participation. If you are the last one in the room, volunteer to help them take down the testing location. They will appreciate your offer.

And now wait patiently outside for the examiners to announce you have passed both examinations. Chances are they will greet you with a smile and your certificate of completion. Make sure to immediately sign this certificate of completion when it is handed to you.

YOUR NEXT STEP

Did you try the code test? If not, your next step after Technician is Technician Plus. This only requires the 5-wpm code test. As mentioned previously, Radio Shack has the *New Novice Voice Class* which contains an excellent set of code training tapes for learning the code, and passing the test.

It is recommended that you also begin thinking about General Class. If you decide to go ahead and learn code, go ahead and shoot for the stars, and prepare for passing the code test at 13-wpm General Class speed. As mentioned previously, Radio Shack also has license preparation materials, *New General Class,* that builds code speeds from 5 to 13 wpm. The Radio Shack product includes a Gordon West authored book with the Element 3B question pool to make it a breeze to pass the 25-question General Class written examination and the Element 1B 13-wpm code test.

So don't stop now at Technician or Technician-Plus—go for General Class! That's your next step.

TECHNICIAN CLASS CALL SIGNS

Technician Class call signs are selected from category Group C by a computer in Gettysburg, Pennsylvania. It's a sequential selection, so sorry, you can't pick out a favorite. Your call letters will consist of a single letter, a single number, followed by 3 letters (N6NOA). Unfortunately, there are some parts of the country where all of the Group C call letters have been used up. This means you may receive a Group D set of Technician call letters, beginning with 2 letters, a number, and followed by 3 letters (KB1XYZ).

But whatever call signs you get, you will be proud of them because you are the only one in the world with that particular FCC-issued call sign.

When you upgrade to Advanced Class, you have another opportunity to change call signs.

CONGRATULATIONS! YOU PASSED

After you pass the examinations, congratulations are in order and a big welcome to Technician Class privileges. The world of microwave and VHF/UHF operating awaits you. And, if you also passed the code test, welcome to Technician Plus privileges and the world of long-range high-frequency operation.

You cannot begin operating until your official FCC call sign arrives. It will take from 6 to 8 weeks for your call sign to be processed, and sent to the address listed on your Form 610. You can buy equipment ahead of time, but hide the microphone until your ticket arrives in the mail. You cannot go on the air until your own call sign hits town.

SUMMARY

Welcome to the world of Amateur Radio. Maybe I'll hear you on the airwaves soon. Here in Southern California, I'm on the 2-meter band on 144.330 MHz. If you have a Technician-Plus license, you can find me on 10 meters at 28.303 most mornings. And if the band is open, I'm also on the worldwide 6-meter band near 50.120 MHz.

I would also like you to write me and send us a double-stamped, self-addressed envelope for an exclusive Technician Class passing certificate. I'll sign it personally, and I welcome you to the fascinating world of Amateur Radio.

Hope to hear you on the bands soon.

Gordon West, WB6NOA

Appendix

U.S. VOLUNTEER-EXAMINER COORDINATORS IN THE AMATEUR SERVICE

Anchorage Amateur Radio Club
2628 Turnagain Parkway
Anchorage, AK 99503
(907) 243-2221, 344-5401

ARRL/VEC*
225 Main Street
Newington, CT 06111
(203) 666-1541

Central Alabama VEC, Inc.
606 Tremont Street
Selma, AL 36701
H: (205) 872-1166, 875-7419
O: (205) 874-1688

Charlotte VEC
227 Bennett Lane
Charlotte, NC 28213
(704) 596-2168

DeVry Amateur Radio Society*
3300 North Campbell Avenue
Chicago, IL 60618
(312) 929-8500

Golden Empire Amateur Radio Society
P.O. Box 508
Chico, CA 95927

Greater Los Angeles Amateur Radio
    Group
9737 Noble Avenue
Sepulveda, CA 91343
(818) 762-5095, 892-2068

Jefferson Amateur Radio Club
P.O. Box 73665
Metairie, LA 70033

Koolau Amateur Radio Club
45-529 Nakuluai Street
Kaneohe, HI 96744
(808) 235-4132

Laurel Amateur Radio Club, Inc.
P.O. Box 3039
Laurel, MD 20708-0039
(301) 953-1065

Mountain Amateur Radio Club
P.O. Box 234
Cumberland, MD 21502
(304) 289-3576

PHD Amateur Radio Association, Inc.
P.O. Box 11
Liberty, MO 64068
(816) 781-7313

Triad Emergency Amateur Radio Club
3504 Stonehurst Place
High Point, NC 27260
(919) 841-7576

Sandarc-VEC
P.O. Box 2456
La Mesa, CA 92044
(619) 465-3926

Sunnyvale VEC Amateur Radio Club
P.O. Box 60142
Sunnyvale, CA 94088-0142
(408) 255-9000

The Milwaukee Radio Amateurs
    Club, Inc.
1737 N. 116th St.
Wauwatosa, WI 53226
(414) 774-6999

Western Carolina Amateur
    Radio Society
5833 Clinton Hwy, Suite 203
Knoxville, TN 37912-2545
(615) 688-7771

W5YI-VEC*
P.O. Box 565101
Dallas, TX 75356-5101
(817) 461-6443

*This VEC regularly offers monthly exams
in all parts of the country.

Write for graduation certificates to:
    Gordon West's Radio School
    2414 College Drive
    Costa Mesa, CA 92626

Please include:
    1. Certificate of completion
    2. One double-stamped self-addressed
       envelope

AUTHORIZED FREQUENCY BANDS – AMATEUR SERVICE
(for U.S. Amateur Stations operating from ITU-Region 2–North and South America)

Meters	Novice	Technician[1,2]	Technician Plus[2]	General	Advanced	Extra Class
160				1800-2000 kHz/All	1800-2000 kHz/All	1800-2000 kHz/All
80	3675-3725 kHz/CW		3675-3725 kHz/CW	3525-3750 kHz/CW 3850-4000 kHz/Ph	3525-3750 kHz/CW 3775-4000 kHz/Ph	3500-4000 kHz/CW 3750-4000 kHz/Ph
40	7100-7150 kHz/CW		7100-7150 kHz/CW	7025-7150 kHz/CW 7225-7300 kHz/Ph	7025-7300 kHz/CW 7150-7300 kHz/Ph	7000-7300 kHz/CW 7150-7300 kHz/Ph
30				10.1-10.15 MHz/CW	10.1-10.15 MHz/CW	10.1-10.15 MHz/CW
20				14.025-14.15 MHz/CW 14.225-14.35 MHz/Ph	14.025-14.15 MHz/CW 14.175-14.35 MHz/Ph	14.0-14.35 MHz/CW 14.15-14.35 MHz/Ph
15	21.1-21.2 MHz/CW		21.1-21.2 MHz/CW	21.025-21.2 MHz/CW 21.3-21.45 MHz/Ph	21.025-21.2 MHz/CW 21.225-21.45 MHz/Ph	21.0-21.45 MHz/CW 21.2-21.45 MHz/Ph
12				24.89-24.99 MHz/CW 24.93-24.99 MHz/Ph	24.89-24.99 MHz/CW 24.93-24.99 MHz/Ph	24.89-24.99 MHz/CW 24.93-24.99 MHz/Ph
10	28.1-28.5 MHz/CW 28.3-28.5 MHz/Ph		28.1-28.5 MHz/CW 28.3-28.5 MHz/Ph	28.0-29.7 MHz/CW 28.3-29.7 MHz/Ph	28.0-29.7 MHz/CW 28.3-29.7 MHz/Ph	28.0-29.7 MHz/CW 28.3-29.7 MHz/Ph
6		50-54 MHz/CW 50.1-54 MHz/Ph	50-54 MHz/CW 50.1-54 MHz/Ph	50-54 MHz/CW 50.1-54 MHz/Ph	50-54 MHz/CW 50.1-54 MHz/Ph	50-54 MHz/CW 50.1-54 MHz/Ph
2		144-148 MHz/CW 144.1-148 MHz/All	144-148 MHz/CW 144.1-148 MHz/All	144-148 MHz/CW 144.1-148 MHz/Ph	144-148 MHz/CW 144.1-148 MHz/All	144-148 MHz/CW 144.1-148 MHz/All
*1.25	222.1-223.91 MHz/All	222-225 MHz/All	222-225 MHz/All	222-225 MHz/All	222-225 MHz/All	222-225 MHz/All
0.70		420-450 MHz/All	420-450 MHz/All	420-450 MHz/All	420-450 MHz/All	420-450 MHz/All
0.35		902-928 MHz/All	902-928 MHz/All	902-928 MHz/All	902-928 MHz/All	902-928 MHz/All
0.23	1270-1295 MHz/All	1240-1300 MHz/All	1240-1300 MHz/All	1240-1300 MHz/All	1240-1300 MHz/All	1240-1300 MHz/All

[1] No-Code License   [2] Effective 2/14/91

Note: Morse code (CW, A1A) may be used on any frequency allocated to the amateur service. Telephony emission (abbreviated Ph above) authorized on certain bands as indicated. Higher class licensees may use slow-scan television and facsimile emissions on the Phone bands; radio teletype/digital on the CW bands. All amateur modes and emissions are authorized above 144.1 MHz. In actual practice, the modes/emissions used are somewhat more complicated than shown above due to the existence of various band plans and "gentlemen's agreements" concerning where certain operations should take place.

*FCC has ruled exclusive allocation for Amateur Radio of 222-225 MHz effective1/1/90. The 220-222 MHz has been allocated to Land Mobile. Operation on these frequencies by amateurs will cease.

ITU REGIONS

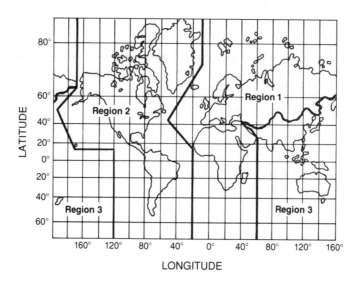

US CALL SIGNS

§ PART 97.3 FCC EMISSION TERMS

(1) CW. International Morse code telegraphy emissions having designators with A, C, H, J or R as the first symbol; 1 as the second symbol; A or B as the third symbol; and emissions J2A and J2B.

(2) Data. Telemetry, telecommand and computer communications emissions having designators with A, C, D, F, G, H, J or R as the first symbol; 1 as the second symbol; D as the third symbol; and emission J2D. Only a digital code of a type specifically authorized in this Part may be transmitted.

(3) Image. Facsimile and television emissions having designators with A, C, D, F, G, H, J or R as the first symbol; 1, 2 or 3 as the second symbol; C or F as the third symbol; and emissions having B as the first symbol; 7, 8 or 9 as the second symbol; W as the third symbol.

(4) MCW. Tone-modulated international Morse code telegraphy emissions having designators with A, C, D, F, G, H or R as the first symbol; 2 as the second symbol; A or B as the third symbol.

(5) Phone. Speech and other sound emissions having designators with A, C, D, F, G, H, J or R as the first symbol; 1, 2 or 3 as the second symbol; E as the third symbol. Also speech emissions having B as the first symbol; 7, 8 or 9 as the second symbol; E as the third symbol. MCW for the purpose of performing the station identification procedure, or for providing telegraphy practice interspersed with speech. Incidental tones for the purpose of selective calling or alerting or to control the level of a demodulated signal may also be considered phone.

(6) Pulse. Emissions having designators with K, L, M, P, Q, V or W as the first symbol; 0, 1, 2, 3, 7, 8, 9 or X as the second symbol; A, B, C, D, E, F, N, W or X as the third symbol.

(7) RTTY. Narrow-band direct-printing telegraphy emissions having designators with A, C, D, F, G, H, J or R as the first symbol; 1 as the second symbol; B as the third symbol; and emission J2B. Only a digital code of a type specifically authorized in this Part may be transmitted.

(8) SS. Spread-spectrum emissions using bandwidth-expansion modulation emissions having designators with A, C, D, F, G, H, J or R as the first symbol; X as the second symbol; X as the third symbol. Only a SS emission of a type specifically authorized in this Part may be transmitted.

(9) Test. Emissions containing no information having the designators with N as the third symbol. Test does not include pulse emission with no information or modulation unless pulse emissions are also authorized in the frequency band.

ITU EMISSION DESIGNATORS

First symbol – type of main carrier modulation
(Note: Modulation is the process of varying the radio wave to convey information.)
- N – Emission of an unmodulated carrier
- A – Amplitude Modulation
- J – Single sideband, suppressed carrier (the emission you are authorized on ten meters between 28.3 and 28.5 MHz)
- F – Frequency modulation
- G – Phase modulation
- P – Sequence of unmodulated pulses

Second symbol – Nature of signals modulating the main carrier
- $\emptyset$ – No modulating signal
- 1 – Single channel digital information, no modulation
- 2 – Single channel digital information with modulation
- 3 – Single channel carrying analog information
- 7 – Two or more channels carrying digital information
- 8 – Two or more channels carrying analog information
- 9 – Combination of digital and analog information

Third symbol – Type of information to be transmitted
- N – No information transmitted
- A – Telegraphy to be received by human hearing
- B – Telegraphy to be received by automatic equipment
- C – Facsimile, transmission of pictures by radio
- D – Data transmission, telemetry, telecommand
- E – Telephony (voice information)
- F – Video, television
- W – Combination of above

Amplitude Modulated	Traditional Symbol	New Symbol
Unmodulated	A$\emptyset$	N$\emptyset$N
Keyed on/off	A1	A1A
Tones keyed on/off	A2	A2A
AM data		A2D
Keyed tones w/SSB	A2J	J2A
SSB data		J2D
AM voice	A3	A3E
Voice w/SSB	A3J	J3E
AM facimile	A4	A3C
SSB television	A5	C3F
AM television	A5	A3F

Source: FCC

Frequency Modulated	Traditional Symbol	New Symbol
Unmodulated	F$\emptyset$	N$\emptyset$N
Switched between two frequencies	F1	F1B
Switch tones	F2	F2A
FM data		F2D
FM voice	F3	F3E
FM facsimile	F4	F3C
FM television	F5	F3F

Pulse Modulated		
Phase	P	P1B

LIST OF COUNTRIES PERMITTING THIRD-PARTY TRAFFIC

V2	Antigua & Barbuda	YS	El Salvador	ZP	Paraguay
LU	Argentina	C5	The Gambia	OA	Peru
VK	Australia	9G	Ghana	VR6	Pitcairn Island*
V3	Belize	J3	Grenada	V4	St Christopher (St. Kitts)
CP	Bolivia	TG	Guatemala		and Nevis
PY	Brazil	8R	Guyana	J6	St. Lucia
VE	Canada	HH	Haiti	J8	St. Vincent
CE	Chile	HR	Honduras	3D6	Swaziland
HK	Colombia	4X	Israel	9Y	Trinidad & Tobago
TI	Costa Rica	6Y	Jamaica	GB	United Kingdom**
CO	Cuba	JY	Jordan	CX	Uruguay
J7	Dominica	EL	Liberia	YV	Venezuela
HI	Dominican Rep.	XE	Mexico	4U1ITU	ITU, Geneva
HC	Ecuador	YN	Nicaragua	4U1VIC	VIC, Geneva
		HP	Panama		

* Informal Temporary
** Limited to special-event stations with callsign prefix GB; GB3 excluded   4/91

FOR NOVICE TESTING

For the applicant that knows some code and wants to enter amateur services through the Novice route, a summary of instructions to the applicant and the two General Class or higher operators that want to administer the Novice examination is contained in the following: (Source: FCC)

INSTRUCTIONS TO APPLICANTS FOR NOVICE OPERATOR LICENSE

1. The examination for the Novice operator license is administered by two eligible VEs of your choice. The examination must be administered at a place and time specified by the administering VEs.
2. Complete Section I of the FCC Form 610 and present it to the VEs.
3. Bring to the examination session and show to the VEs:
   A. At least two documents to prove your identity to the VEs;
   B. The original document of your Certificate of Successful Completion of Examination (if you have one), if issued to you within 365 days of this examination session;
   C. The original document of your FCC Commercial Radiotelegraph Operator license (if you have one) if it is current or was current within five years prior to the examination session.
4. Your VEs will administer Element 1(A), unless you have a valid Certificate of Successful Completion of Examination for Elements 1(A), 1(B), or 1(C), or unless you have an FCC Commercial Radiotelegraph Operator license.
5. Your VEs will administer Element 2, unless you have a valid Certificate of Successful Completion of Examination for Element 2.

INSTRUCTIONS TO TWO HAMS ADMINISTERING
THE NOVICE OPERATOR LICENSE

1. Review Section I of the examinee's FCC Form 610. Have the examinee make any necessary corrections.
2. Review the examinee's identification documents. Satisfy yourselves that the person to who you administer the examination is the person applying on the FCC Form 610.
3. If the examinee has a valid Certificate of Successful Completion of Examination issued within the previous 365 days, enter the date of issuance of the Certificate in the appropriate box in ITEM B of the Administering VEs' Report.
4. If the examinee has a valid (or expired less than 5 years prior to the examination session) FCC Commercial Radiotelegraph Operator license, enter the license number and the expiration date in the appropriate boxes in ITEM C of the Administering VEs' Report.
5. Administer the required element(s) (Element 1(A), Element 2, or both). For the element(s) you administer which the examinee passes, make a check mark in the appropriate box(es) in ITEM D of the Administering VEs' Report.
6. If the examinee qualifies for the Novice operator license:
   A. Make a check mark in the box in ITEM E1 of the Administering VEs' Report.
   B. Each VE must complete Section II-A:
      1. One VE must complete ITEMS 1A, 1B, 1C, 1D, 1E, 1F, and 1G.
      2. The second VE must complete ITEMS 2A, 2B, 2C, 2D, 2E, 2F, and 2G.
   C. A photocopy of the completed Form 610 should be made for the successful applicant to present to the Administering VEs in Technician Plus upgrade attempt at an examination session while the application is pending.
   D. Within ten days of administering the examination, send the FCC Form 610 to: Federal Communications Commission, Gettysburg, PA 17326.
7. If the examinee does not qualify for the Novice operator license, but passes Element 1A or 2:
   A. Make a check mark in the "None" box in ITEM E of the Administering VEs' Report.
   B. Make a check mark in the appropriate box in ITEM D of the Administering VEs' Report.
   C. Each administering VE must complete ITEMS 1A, 1B, 1C, 1D, 1E, 1F, and 1G.
   D. Return the application to the unsuccessful examinee.
8. If the examinee does not qualify for the Novice operator license, return the application form to the examinee and inform the examinee of the grade.

COMMON CW ABBREVIATIONS

AA	All after	NR	Number
AB	All before	NW	Now; I resume transmission
ABT	About		
ADR	Address	OB	Old boy
AGN	Again	OM	Old man
ANT	Antenna	OP-OPR	Operator
BCI	Broadcast interference	OT	Old timer; old top
BK	Break; break me; break in	PBL	Preable
BN	All between; been	PSE-PLS	Please
B4	Before	PWR	Power
C	Yes	PX	Press
CFM	Confirm; I confirm	R	Received as transmitted; are
CK	Check		
CL	I am closing my station; call	RCD	Received
		REF	Refer to; referring to; reference
CLD-CLG	Called; calling		
CUD	Could	RPT	Repeat; I repeat
CUL	See you later	SED	Said
CUM	Come	SEZ	Says
CW	Continuous Wave	SIG	Signature; signal
DLD-DLVD	Delivered	SKED	Schedule
DX	Distance	SRI	Sorry
FB	Fine business; excellent	SVC	Service; prefix to service message
GA	Go ahead (or resume sending)		
		TFC	Traffic
GB	Good-by	TMW	Tomorrow
GBA	Give better address	TNX	Thanks
GE	Good evening	TU	Thank you
GG	Going	TVI	Television interference
GM	Good morning	TXT	Text
GN	Good night	UR-URS	Your; you're; yours
GND	Ground	VFO-	Variable-frequency oscillator
GUD	Good		
HI	The telegraphic laugh; high	VY	Very
		WA	Word after
HR	Here; hear	WB	Word before
HV	Have	WD-WDS	Word; words
HW	How	WKD-WKG	Worked; working
LID	A poor operator	WL	Well; will
MILS	Milliamperes	WUD	Would
MSG	Message; prefix to radiogram	WX	Weather
		XMTR	Transmitter
N	No	XTAL	Crystal
ND	Nothing doing	XYL	Wife
NIL	Nothing; I have nothing for you	YL	Young lady
		73	Best regards
		88	Love and kisses

POPULAR Q SIGNALS

Given below are a number of Q signals whose meanings most often need to be expressed with brevity and clarity in amateur work. (Q abbreviations take the form of questions only when each is sent followed by a question mark.)

QRG Will you tell me my exact frequency (or that of _____)? Your exact frequency (or that of _____) is _____ kHz.

QRH Does my frequency vary? Your frequency varies.

QRI How is the tone of my transmission? The tone of your transmission is _____ (1. Good; 2. Variable; 3. Bad).

QRJ Are you receiving me badly? I cannot receive you. Your signals are too weak.

QRK What is the intelligibility of my signals (or those of _____)? The intelligibility of your signals (or those of _____) is _____ (1. Bad; 2. Poor; 3. Fair; 4. Good; 5. Excellent).

QRL Are you busy? I am busy (or I am busy with _____). Please do not interfere.

QRM Is my transmission being interfered with? Your transmission is being interfered with _____ (1. Nil; 2. Slightly; 3. Moderately; 4. Severely; 5. Extremely).

QRN Are you troubled by static? I am troubled by static _____ (1-5 as under QRM).

QRO Shall I increase power? Increase power.

QRP Shall I decrease power? Decrease power.

QRQ Shall I send faster? Send faster (_____ WPM).

QRS Shall I send more slowly? Send more slowly (_____ WPM).

QRT Shall I stop sending? Stop sending.

QRU Have you anything for me? I have nothing for you.

QRV Are you ready? I am ready.

QRW Shall I inform _____ that you are calling on _____ kHz? Please inform _____ that I am calling on _____ kHz.

QRX When will you call me again? I will call you again at _____ hours (on _____ kHz).

QRY What is my turn? Your turn is numbered _____.

QRZ Who is calling me? You are being called by _____ (on _____ kHz).

QSA What is the strength of my signals (or those of _____)? The strength of your signals (or those of _____) is _____ (1. Scarcely perceptible; 2. Weak; 3. Fairly good; 4. Good; 5. Very good).

QSB Are my signals fading? Your signals are fading.

QSD Is my keying defective? Your keying is defective.

QSG Shall I send _____ messages at a time? Send _____ messages at a time.

QSK Can you hear me between your signals and if so can I break in on your transmission? I can hear you between my signals; break in on my transmission.

QSL Can you acknowledge receipt? I am acknowledging receipt.

QSM Shall I repeat the last message which I sent you, or some previous message? Repeat the last message which you sent me [or message(s) number(s) _____].

QSN Did you hear me (or ____) on ____ kHz? I heard you (or ____) on ____ kHz.

QSO Can you communicate with ____ direct or by relay? I can communicate with ____ direct (or by relay through ____).

QSP Will you relay to ____? I will relay to ____.

QST General call preceding a message addressed to all amateurs and ARRL members. This is in effect "CQ ARRL."

QSU Shall I send or reply on this frequency (or on ____ kHz)?

QSW Will you send on this frequency (or on ____ kHz)? I am going to send on this frequency (or on ____ kHz).

QSX Will you listen to ____ on ____ kHz? I am listening to ____ on ____ kHz.

QSY Shall I change to transmission on another frequency? Change to transmission on another frequency (or on ____ kHz).

QSZ Shall I send each word or group more than once? Send each word or group twice (or ____ times).

QTA Shall I cancel message number ____? Cancel message number ____.

QTB Do you agree with my counting of words? I do not agree with your counting of words. I will repeat the first letter or digit of each word or group.

QTC How many messages have you to send? I have messages for you (or for ____).

QTH What is your location? My location is ____.

QTR What is the correct time? The time is ____.

Source: ARRL

COUNTRIES HOLDING U.S. RECIPROCAL AGREEMENTS

Antigua	Germany (Fed. Rep.)	Netherlands Ant.
Argentina	Greece	New Zealand
Australia	Grenada	Nicaragua
Austria	Guatemala	Norway
Bahamas	Guyana	Panama
Barbados	Haita	Paraguay
Belgium	Honduras	Peru
Belize	Hong Kong	Phillippineas
Bolivia	Iceland	Portugal
Botswana	India	St. Lucia
Brazil	Indonesia	Sierra Leone
Canada	Ireland	Solomon Islands
Chile	Israel	South Africa
Colombia	Italy	Spain
Costa Rica	Jamaica	Suriname
Cyprus	Japan	Sweden
Denmark	Jordan	Switzerland
Dominica (Comwth)	Kiribati	Trinidad
Dominican Rep.	Kuwait	Tuvalu
Ecuador	Liberia	United Kingdom
El Salvador	Luxembourg	Uruguay
Fiji	Monaco	Venezuela
Finland	Netherlands	Yugoslavia
France		

4/91

FEDERAL COMMUNICATIONS COMMISSION
GETTYSBURG, PA 17326

Approved OMB
3060-0003
Expires 12/31/92
See instructions for information
regarding public burden estimate

APPLICATION FOR AMATEUR RADIO
STATION AND/OR OPERATOR LICENSE

ADMINISTERING VEs' REPORT		EXAMINATION ELEMENTS							
Applicant is credited for:		1(A)	1(B)	1(C)	2	3(A)	3(B)	4(A)	4(B)
A. CIRCLE CLASS OF FCC AMATEUR LICENSE HELD: N T G A Class →		(NT)	(GA)		(NTGA)	(TGA)	(GA)	(A)	
B. CERTIFICATE(S) OF SUCCESSFUL COMPLETION OF AN EXAMINATION HELD: →		Date Issued	Date Issued	Date Issued	Date Issued	Date Issued	Date Issued	Date Issued	Date Issued
C. FCC COMMERCIAL RADIOTELEGRAPH OPERATOR LICENSE HELD: Number: Exp. Date:									
D. EXAMINATION ELEMENTS PASSED THAT WERE ADMINISTERED AT THIS SESSION: →									

E. APPLICANT IS QUALIFIED FOR OPERATOR LICENSE CLASS: ☐ NONE:

E1. ☐ NOVICE (Elements 1(A), 1(B), or 1(C) and 2)

E2. ☐ TECHNICIAN (Elements 1(A), 1(B), or 1(C), 2 and 3(A))
☐ GENERAL (Elements 1(B) or 1(C), 2, 3(A), and 3(B))
☐ ADVANCED (Elements 1(B) or 1(C), 2, 3(A), 3(B) and 4(A))
☐ AMATEUR EXTRA (Elements 1(C), 2, 3(A), 3(B), 4(A), and 4(B))

H. Date of VEC coordinated examination session:

I. VEC Receipt Date:

F. NAME OF VOLUNTEER-EXAMINER COORDINATOR: (VEC coordinated sessions only)

G. EXAMINATION SESSION LOCATION: (VEC coordinated sessions only)

SECTION I

1. IF YOU HOLD A VALID LICENSE ATTACH THE ORIGINAL LICENSE OR PHOTOCOPY ON BACK OF APPLICATION. IF THE VALID LICENSE OR CERTIFICATE OF SUCCESSFUL COMPLETION OF AN EXAMINATION WAS LOST OR DESTROYED, PLEASE EXPLAIN.

2. CHECK ONE OR MORE ITEMS, NORMALLY ALL LICENSES ARE ISSUED FOR A 10 YEAR TERM.

2A. ☐ RENEW LICENSE – NO OTHER CHANGES → EXPIRATION DATE (Month, Day, Year)

2B. ☐ REINSTATE LICENSE EXPIRED LESS THAN 2 YEARS →

2C. ☑ EXAMINATION FOR NEW LICENSE

2D. ☐ EXAMINATION TO UPGRADE OPERATOR CLASS FORMER LAST NAME   SUFFIX (Jr., Sr., etc.)

2E. ☐ CHANGE CALL SIGN (Be sure you are eligible – See Inst. 2E)

2F. ☐ CHANGE NAME (Give former name) →   FORMER FIRST NAME   MIDDLE INITIAL

2G. ☐ CHANGE MAILING ADDRESS

2H. ☐ CHANGE STATION LOCATION

3. CALL SIGN (If you checked 2C above, skip items 3 and 4)   4. OPERATOR CLASS OF THE ATTACHED LICENSE:

5. CURRENT FIRST NAME **MEECHIE**	M.I. **C**	LAST NAME **WEST**	SUFFIX (Jr., Sr., etc.)	6. DATE OF BIRTH **06/22/72** MONTH DAY YEAR

7. CURRENT MAILING ADDRESS (Number and Street) **12345 FELINE DR.**   CITY **DOGVILLE**   STATE **CA**   ZIP CODE **92001**

8. CURRENT STATION LOCATION (Do not use a P.O. Box No., RFD No., or General Delivery. See Instruction 8) **12345 FELINE DR**   CITY **DOGVILLE**   STATE **CA**

9. Would a Commission grant of your application be an action which may have a significant environmental effect as defined by Section 1.1307 of the Commission's Rules? See instruction 9. If you answer yes, submit the statement as required by Sections 1.1308 and 1.1311. ☐ YES ☑ NO

10. Do you have any other amateur radio application on file with the Commission that has not been acted upon? If yes, answer items 11 and 12. ☐ YES ☑ NO

11. PURPOSE OF OTHER APPLICATION   12. DATE SUBMITTED (Month, Day, Year)

CERTIFICATION

I CERTIFY THAT all statements herein and attachments herewith are true, complete, and correct to the best of my knowledge and belief and are made in good faith; that I am not a representative of a foreign government; that I waive any claim to the use of any particular frequency regardless of prior use by license or otherwise; and that the station to be licensed will be inaccessible to unauthorized persons.

WILLFUL FALSE STATEMENTS MADE ON THIS FORM OR ATTACHMENTS ARE PUNISHABLE BY FINE AND IMPRISONMENT
U.S. CODE TITLE 18, SECTION 1001

13. SIGNATURE OF APPLICANT: (Must match Item 5) *Meechie C. West*   14. DATE SIGNED: **02-14-91**

(OVER)   FCC Form 610, February 1990

(Front Side – Filled-in area for applicant, shaded area for VEs)

Form 610 – Technician Class License Application Form

FEDERAL COMMUNICATIONS COMMISSION
GETTYSBURG, PA 17326

**APPLICATION FOR AMATEUR RADIO
STATION AND/OR OPERATOR LICENSE**

Approved OMB
3060-0003
Expires 12/31/92
See instructions for information
regarding public burden estimate

ADMINISTERING VEs' REPORT		EXAMINATION ELEMENTS							
Applicant is credited for: ➡		1(A)	1(B)	1(C)	2	3(A)	3(B)	4(A)	4(B)
A. CIRCLE CLASS OF FCC AMATEUR LICENSE HELD: N T G A **Class**		(NT)	(GA)	(1C)	(NTGA)	(TGA)	(GA)	(A)	
B. CERTIFICATE(S) OF SUCCESSFUL COMPLETION OF AN EXAMINATION HELD: ➡		Date Issued	Date Issued	Date Issued	Date Issued	Date Issued	Date Issued	Date Issued	Date Issued
C. FCC COMMERCIAL RADIOTELEGRAPH OPERATOR LICENSE HELD:	Number: Exp. Date:								
D. EXAMINATION ELEMENTS PASSED THAT WERE ADMINISTERED AT THIS SESSION: ➡									

E. APPLICANT IS QUALIFIED FOR OPERATOR LICENSE CLASS: ☐ NONE:

E1. ☐ NOVICE   (Elements 1(A), 1(B), or 1(C) and 2)
E2. ☐ TECHNICIAN   (Elements 1(A), 1(B), or 1(C). 2 and 3(A))
☐ GENERAL   (Elements 1(B) or 1(C). 2, 3(A), and 3(B))
☐ ADVANCED   (Elements 1(B) or 1(C). 2, 3(A), 3(B) and 4(A))
☐ AMATEUR EXTRA   (Elements 1(C). 2, 3(A). 3(B). 4(A). and 4(B))

H. Date of VEC coordinated examination session:

I. VEC Receipt Date:

F. NAME OF VOLUNTEER-EXAMINER COORDINATOR: (VEC coordinated sessions only)

G. EXAMINATION SESSION LOCATION: (VEC coordinated sessions only)

SECTION I

1. IF YOU HOLD A VALID LICENSE ATTACH THE ORIGINAL LICENSE OR PHOTOCOPY ON BACK OF APPLICATION. IF THE VALID LICENSE OR CERTIFICATE OF SUCCESSFUL COMPLETION OF AN EXAMINATION WAS LOST OR DESTROYED, PLEASE EXPLAIN.

2. CHECK ONE OR MORE ITEMS, NORMALLY ALL LICENSES ARE ISSUED FOR A 10 YEAR TERM.

2A. ☐ RENEW LICENSE – NO OTHER CHANGES ➡   EXPIRATION DATE (Month, Day, Year)
2B. ☐ REINSTATE LICENSE EXPIRED LESS THAN 2 YEARS ➡
2C. ☐ EXAMINATION FOR NEW LICENSE
2D. ☒ EXAMINATION TO UPGRADE OPERATOR CLASS   FORMER LAST NAME   SUFFIX (Jr., Sr., etc.)
2E. ☒ CHANGE CALL SIGN (Be sure you are eligible – See Inst. 2E)
2F. ☐ CHANGE NAME (Give former name) ➡   FORMER FIRST NAME   MIDDLE INITIAL
2G. ☐ CHANGE MAILING ADDRESS
2H. ☐ CHANGE STATION LOCATION

3. CALL SIGN (If you checked 2C above, skip items 3 and 4)
WA6BWH

4. OPERATOR CLASS OF THE ATTACHED LICENSE:
TECHNICIAN

5. CURRENT FIRST NAME   MARGARET   **M.I.** E.   **LAST NAME** FOSTER   SUFFIX (Jr., Sr., etc.)   **6.** DATE OF BIRTH 02/16/42   MONTH DAY YEAR

7. CURRENT MAILING ADDRESS (Number and Street)   12345 MICHI STREET   **CITY** CATSVILLE   **STATE** CA.   **ZIP CODE** 92626

8. CURRENT STATION LOCATION (Do not use a P.O. Box No., RFD No., or General Delivery. See Instruction 8)   12345 MICHI STREET   **CITY** CATSVILLE   **STATE** CA.

9. Would a Commission grant of your application be an action which may have a significant environmental effect as defined by Section 1.1307 of the Commission's Rules? See instruction 9. If you answer yes, submit the statement as required by Sections 1.1308 and 1.1311. ☐ YES ☒ NO

10. Do you have any other amateur radio application on file with the Commission that has not been acted upon? If yes, answer items 11 and 12. ☐ YES ☒ NO

11. PURPOSE OF OTHER APPLICATION

12. DATE SUBMITTED (Month, Day, Year)

CERTIFICATION

I CERTIFY THAT all statements herein and attachments herewith are true, complete, and correct to the best of my knowledge and belief and are made in good faith; that I am not a representative of a foreign government; that I waive any claim to the use of any particular frequency regardless of prior use by license or otherwise; and that the station to be licensed will be inaccessible to unauthorized persons.

**WILLFUL FALSE STATEMENTS MADE ON THIS FORM OR ATTACHMENTS ARE PUNISHABLE BY FINE AND IMPRISONMENT
U.S. CODE TITLE 18, SECTION 1001**

13. SIGNATURE OF APPLICANT: (Must match Item 5)   Margaret E. Foster   (OVER)

14. DATE SIGNED   2-14-91

FCC Form 610, February 1990

(Front Side – Filled-in area for applicant, shaded area for VEs)

Form 610 – Technician Plus Class License Application Form

ATTACH THE ORIGINAL LICENSE OR PHOTOCOPY HERE

SECTION II – EXAMINATION INFORMATION

SECTION II-A FOR NOVICE OPERATOR EXAMINATION ONLY. To be completed by the Administering VEs after completing the Administering VE's Report on the other side of this form.

CERTIFICATION

I CERTIFY THAT I have complied with the Administering VE requirements stated in Part 97 of the Commission's Rules; THAT I have administered to the applicant and graded an amateur radio operator examination in accordance with Part 97 of the Commission's Rules; THAT I have indicated in the Administering VE's Report the examination element(s) the applicant passed; THAT I have examined documents held by the applicant and I have indicated in the Administering VE's Report the examination element for which the applicant is given examination credit in accordance with Part 97 of the Commission's Rules.

1A. VOLUNTEER EXAMINER'S NAME: (First, MI, Last, Suffix) *(Print or Type)*

1B. VE'S MAILING ADDRESS: (Number, Street, City, State, ZIP Code)

1C. VE'S OPERATOR CLASS: ☐ GENERAL ☐ ADVANCED ☐ AMATEUR EXTRA	1D. VE'S STATION CALL SIGN
1E. LICENSE EXPIRATION DATE:	1F. IF YOU HAVE AN APPLICATION PENDING FOR YOUR LICENSE, GIVE FILING DATE:
1G. SIGNATURE: (Must match Item 1A)	DATE SIGNED

2A. VOLUNTEER EXAMINER'S NAME: (First, MI, Last, Suffix) *(Print or Type)*

2B. VE'S MAILING ADDRESS: (Number, Street, City, State, ZIP Code)

2C. VE'S OPERATOR CLASS: ☐ GENERAL ☐ ADVANCED ☐ AMATEUR EXTRA	2D. VE'S STATION CALL SIGN
2E. LICENSE EXPIRATION DATE:	2F. IF YOU HAVE AN APPLICATION PENDING FOR YOUR LICENSE, GIVE FILING DATE:
2G. SIGNATURE: (Must match Item 2A)	DATE SIGNED

SECTION II-B FOR TECHNICIAN, GENERAL, ADVANCED, OR AMATEUR EXTRA OPERATOR EXAMINATION ONLY. To be completed by the Administering VEs after completing the Administering VE's Report on the other side of this form.

CERTIFICATION

I CERTIFY THAT I have complied with the Administering VE requirements stated in Part 97 of the Commission's Rules; THAT I have administered to the applicant and graded an amateur radio operator examination in accordance with Part 97 of the Commission's Rules; THAT I have indicated in the Administering VE's Report the examination element(s) the applicant passed; THAT I have examined documents held by the applicant and I have indicated in the Administering VE's Report the examination element(s) for which the applicant is given examination credit in accordance with Part 97 of the Commission's Rules.

1A. VOLUNTEER EXAMINER'S NAME: (First, MI, Last, Suffix) *(Print or Type)*	1B. VE'S STATION CALL SIGN:
1C. SIGNATURE: (Must match Item 1A)	DATE SIGNED:
2A. VOLUNTEER EXAMINER'S NAME: (First, MI, Last, Suffix) *(Print or Type)*	2B. VE'S STATION CALL SIGN:
2C. SIGNATURE: (Must match Item 2A)	DATE SIGNED:
3A. VOLUNTEER EXAMINER'S NAME: (First, MI, Last, Suffix) *(Print or Type)*	3B. VE'S STATION CALL SIGN:
3C. SIGNATURE: (Must match Item 3A)	DATE SIGNED:

FCC Form 610
February 1990

(Back Side – Shaded area for VEs)

Form 610 – Technician Class License Application Form

Glossary

Amateur communication: Non-commercial radio communication by or among amateur stations solely with a personal aim and without personal or business interest.

Amateur operator/primary station license: An instrument of authorization issued by the Federal Communications Commission comprised of a station license, and also incorporating an operator license indicating the class of privileges.

Amateur operator: A person holding a valid license to operate an amateur station issued by the Federal Communications Commission. Amateur operators are frequently referred to as ham operators.

Amateur Radio services: The amateur service, the amateur-satellite service and the radio amateur civil emergency service.

Amateur-satellite service: A radiocommunication service using stations on Earth satellites for the same purpose as those of the amateur service.

Amateur service: A radiocommunication service for the purpose of self-training, intercommunication and technical investigations carried out by amateurs; that is, duly authorized persons interested in radio technique solely with a personal aim and without pecuniary interest.

Amateur station: A station licensed in the amateur service embracing necessary apparatus at a particular location used for amateur communication.

AMSAT: Radio Amateur Satellite Corporation, a non-profit scientific organization. (P.O. Box #27, Washington, DC 20044)

ARES: The emergency division of the American Radio Relay League. See RACES

ARRL: American Radio Relay League, national organization of U.S. Amateur Radio operators. (225 Main Street, Newington, CT 06111)

Audio Frequency (AF): The range of frequencies that can be heard by the human ear, generally 20 hertz to 20 kilohertz.

Automatic control: The use of devices and procedures for station control without the control operator being present at the control point when the station is transmitting.

Automatic Volume Control (AVC): A circuit that continually maintains a constant audio output volume in spite of deviations in input signal strength.

Beam or Yagi antenna: An antenna array that receives or transmits RF energy in a particular direction. Usually rotatable.

Block diagram: A simplified outline of an electronic system where circuits or components are shown as boxes.

Broadcasting: Information or programming transmitted by radio means intended for the general public.

Business communications: Any transmission or communication the purpose of which is to facilitate the regular business or commercial affairs of any party. Business communications are prohibited in the amateur service.

Call Book: A published list of all licensed amateur operators available in North American and Foreign editions.

Call sign assignment: The FCC systematically assigns each amateur station their primary call sign. The FCC will not grant a request for a specific call sign.

Certificate of Successful Completion of Examination (CSCE): A certificate of successful completion allowing examination credit for 365 days. Both written and code credit can be authorized.

Coaxial cable, Coax: A concentric two-conductor cable in which one conductor surrounds the other, separated by an insulator.

Control operator: An amateur operator designated by the licensee of an amateur station to be responsible for the station transmissions.

Coordinated repeater station: An amateur repeater station for which the transmitting and receiving frequencies have been recommended by the recognized repeater coordinator.

Coordinated Universal Time (UTC): Sometimes referred to as Greenwich Mean Time, UCT or Zulu time. The time at the zero-degree (0°) Meridian which passes through Greenwich, England. A universal time among all amateur operators.

Crystal: A quartz or similar material which has been ground to produce natural vibrations of a specific frequency. Quartz crystals produce a high degree of frequency stability in radio transmitters.

CW: Continuous wave, another term for the International Morse code.

Dipole antenna: The most common wire antenna. Length is equal to one-half of the wavelength. Fed by coaxial cable.

Dummy antenna: A device or resistor which serves as a transmitter's antenna without radiating radio waves. Generally used to tune up a radio transmitter.

Duplexer: A device that allows a single antenna to be simultaneously used for both reception and transmission.

Effective Radiated Power (ERP): The product of the transmitter (peak envelope) power, expressed in watts, delivered to the antenna, and the relative gain of an antenna over that of a half wave dipole antenna.

Emergency communication: Any amateur communication directly relating to the immediate safety of life of individuals or the immediate protection of property.

Examination Credit Certificate: (See Certificate of Successful Completion of Examination)

Examination Element: Novices are required to pass Element 1(A): Beginner's Morse code test at five (5) words-per-minute and Element 2: Basic law comprising rules and regulations essential to beginning operation, including sufficient elementary radio theory for the understanding of the rules.

FCC Form 610: The amateur service application form for an amateur operator/primary station license. It is used to apply for a new amateur license or to renew or modify an existing license.

Federal Communications Commission (FCC): A board of five Commissioners, appointed by the President, having the power to regulate wire and radio telecommunications in the United States.

Feedline Transmission line: A system of conductors that connects an antenna to a receiver or transmitter.

Field Day: Annual activity sponsored by the ARRL to demonstrate emergency preparedness of amateur operators.

Filter: A device used to block or reduce alternating currents or signals at certain frequencies while allowing others to pass unimpeded.

Frequency: The number of cycles of alternating current in one second.

Frequency coordinator: An individual or organization recognized by amateur operators eligible to engage in repeater operation which recommends frequencies and other operating and/or technical parameters for amateur repeater operation in order to avoid or minimize potential interferences.

Frequency Modulation (FM): A method of varying a radio carrier wave by causing its frequency to vary in accordance with the information to be conveyed.

Frequency privileges: The transmitting frequency bands available to the various classes of amateur operators. The Novice privileges are listed in Part 97.7(a) of the FCC rules.

Ground: A connection, accidental or intentional, between a device or circuit and the earth or some common body and the earth or some common body serving as the earth.

Ground wave: A radio wave that is propagated near or at the earth's surface.

Handi-Ham system: Amateur organization dedicated to assisting handicapped amateur operators. (3915 Golden Valley Road, Golden Valley, MN 55422)

Harmful interference: Interference which seriously degrades, obstructs or repeatedly interrupts the operation of a radio communication service.

Harmonic: A radio wave that is a multiple of the fundamental frequency. The second harmonic is twice the fundamental frequency, the third harmonic, three times, etc.

Hertz: One complete alternating cycle per second. Named after Heinrich R. Hertz, a German physicist. The number of hertz is the frequency of the audio or radio wave.

High Frequency (HF): The band of frequencies that lie between 3 and 30 Megahertz. It is from these frequencies that radio waves are returned to earth from the ionosphere.

High-Pass filter: A device that allows passage of high frequency signals but attenuates the lower frequencies. When installed on a television set, a high-pass filter allows TV frequencies to pass while blocking lower frequency amateur signals.

Ionosphere: Outer limits of atmosphere from which HF amateur communications signals are returned to earth.

Jamming: The intentional malicious interference with another radio signal.

Key clicks, Chirps: Defective keying of a telegraphy signal sounding like tapping or high varying pitches.

Lid: Amateur slang term for poor radio operator.

Linear amplifier: A device that accurately reproduces a radio wave in magnified form.

Long wire: A horizontal wire antenna that is one wavelength or longer in length.

Low-Pass filter: Device connected to worldwide transmitters that inhibits passage of higher frequencies that cause television interference but does not affect amateur transmissions.

Machine: A ham slang word for an automatic repeater station.

Malicious interference: Willful, intentional jamming of radio transmissions.

MARS: The Military Affiliate Radio System. An organization that coordinates the activities of amateur communications with military radio communications.

Maximum authorized transmitting power: Amateur stations must use no more than the maximum transmitter power necessary to carry out the desired communications. The maximum P.E.P. output power levels authorized Novices are 200 watts in the 80-, 40-, 15- and 10-meter bands, 25 watts in the 220-MHz band, and 5 watts in the 1270-MHz bands.

Maximum usable frequency: The highest frequency that will be returned to earth from the ionosphere.

Medium frequency (MF): The band of frequencies that lies between 300 and 3,000 kHz (3 MHz).

Mobile operation: Radio communications conducted while in motion or during halts at unspecified locations.

Mode: Type of transmission such as voice, teletype, code, television, facsimile.

Modulate: To vary the amplitude, frequency or phase of a radio frequency wave in accordance with the information to be conveyed.

Morse code (see CW): The International Morse code, A1A emission. Interrupted continuous wave communications conducted using a dot-dash code for letters, numbers and operating procedure signs.

Novice operator: An FCC licensed entry-level amateur operator in the amateur service. Novices may operate a transmitter in the following meter wavelength bands: 80, 40, 15, 10, 1.25 and 0.23.

Ohm's law: The basic electrical law explaining the relationship between voltage, current and resistance. The current I in a circuit is equal to the voltage E divided by the resistance R, or I = E/R.

OSCAR: Stands for "Orbiting Satellite Carrying Amateur Radio," the name given to a series of satellites designed and built by amateur operators of several nations.

Oscillator: A device for generating oscillations or vibrations of an audio or radio frequency signal.

Packet radio: A digital method of communicating computer-to-computer. A terminal-node controller makes up the packet of data and directs it to another packet station.

Peak Envelope Power (PEP): 1. The power during one radio frequency cycle at the crest of the modulation envelope, taken under normal operating conditions. 2. The maximum power that can be obtained from a transmitter.

Phone patch: Interconnection of amateur service to the public switched telephone network, and operated by the control operator of the station.

Power supply: A device or circuit that provides the appropriate voltage and current to another device or circuit.

Propagation: The travel of electromagnetic waves or sound waves through a medium.

Q-signals: International three-letter abbreviations beginning with the letter Q used primarily to convey information using the Morse code.

QSL Bureau: An office that bulk processes QSL (radio confirmation) cards for (or from) foreign amateur operators as a postage saving mechanism.

RACES (radio amateur civil emergency service): A radio service using amateur stations for civil defense communications during periods of local, regional, or national civil emergencies.

Radiation: Electromagnetic energy, such as radio waves, traveling forth into space from a transmitter.

Radio Frequency (RF): The range of frequencies over 20 kilohertz that can be propagated through space.

Radio wave: A combination of electric and magnetic fields varying at a radio frequency and traveling through space at the speed of light.

Repeater operation: Automatic amateur stations that retransmit the signals of other amateur stations. Novices may operate through amateur repeaters in the 222-MHz and 1270-MHz bands.

RST Report: A telegraphy signal report system of Readability, Strength and Tone.

S-meter: A voltmeter calibrated from 0 to 9 that indicates the relative signal strength of an incoming signal at a radio receiver.

Selectivity: The ability of a circuit (or radio receiver) to separate the desired signal from those not wanted.

Sensitivity: The ability of a circuit (or radio receiver) to detect a specified input signal.

Short circuit: An unintended low resistance connection across a voltage source resulting in high current and possible damage.

Shortwave: The high frequencies that lie between 3 and 30 Megahertz that are propagated long distances.

Single-Sideband (SSB): A method of radio transmission in which the RF carrier and one of the sidebands is suppressed and all of the information is carried in the one remaining sideband.

Skip wave, Skip zone: A radio wave reflected back to earth. The distance between the radio transmitter and the site of a radio wave's return to earth.

Sky wave: A radio wave that is refracted back to earth in much the same way that a stone thrown across water skips out. Sometimes called an ionospheric wave.

Spectrum: A series of radiated energies arranged in order of wavelength. The radio spectrum extends from 20 kilohertz upward.

Spurious Emissions: Unwanted radio frequency signals emitted from a transmitter that sometimes causes interference.

Station license, location: No transmitting station shall be operated in the amateur service without being licensed by the Federal Communications Commission. Each amateur station shall have one land location, the address of which appears in the station license.

Technician: A no-code amateur operator who has all privileges from 6 meters on up to shorter wavelengths, but *no* privileges on 10, 15, 40, or 80 meter wavelength bands.

Technician Plus: An amateur operator who has passed a 5-wpm code test in addition to Technician Class requirements.

Sunspot Cycle: An 11-year cycle of solar disturbances which greatly affects radio wave propagation.

Telegraphy: Telegraphy is communications transmission and reception using CW International Morse Code.

Telephony: Telephony is communications transmission and reception in the voice mode.

Telecommunications: The electrical conversion, switching, transmission and control of audio signals by wire or radio. Also includes video and data communications.

Temporary operating authority: Authority to operate your amateur station while awaiting arrival of an upgraded license. New Novice operators are not granted temporary operating authority and must await receipt of their new license and call sign.

Terrestrial station location: Any location within the major portion of the earth's atmosphere, including air, sea and land locations.

Third-party traffic: Amateur communication by or under the supervision of the control operator at an amateur station to another amateur station on behalf of others.

Transceiver: A combination radio transmitter and receiver.

Transmatch: An antenna tuner used to match the impedance of the transmitter output to the transmission line of an antenna.

Transmitter: Equipment used to generate radio waves. Most commonly, this radio carrier signal is amplitude varied or frequency varied (modulated) with information and radiated into space.

Transmitter power: The average peak envelope power (output) present at the antenna terminals of the transmitter. The term "transmitted" includes any external radio frequency power amplifier which may be used.

Ultra High Frequency (UHF): Ultra high frequency radio waves that are in the range of 300 to 3,000 MHz.

Upper Sideband (USB): The proper operating mode for sideband transmissions made in the new Novice 10-meter voice band. Amateurs generally operate USB at 20 meters and higher frequencies; lower sideband (LSB) at 40 meters and lower frequencies.

Very High Frequency (VHF): Very high frequency radio waves that are in the range of 30 to 300 MHz.

Volunteer Examiner: An amateur operator of at least a General Class level who administers or prepares amateur operator license examinations. A VE must be at least 18 years old and not related to the applicant.

Volunteer Examiner Coordinator (VEC): A member of an organization which has entered into an agreement with the FCC to coordinate the efforts of volunteer examiners in preparing and administering examinations for amateur operator licenses.

Index

Notes

Notes

Notes

Notes

Notes

Notes